IN PRAISE OF POISON IVY

In Praise of Poison Ivy

The Secret Virtues, Astonishing History, and Dangerous Lore of the World's Most Hated Plant

Anita Sanchez

TAYLOR TRADE PUBLISHING
Lanham • Boulder • New York • London

TAYLOR TRADE PUBLISHING
An imprint of Rowman & Littlefield

Distributed by NATIONAL BOOK NETWORK

British Library Cataloguing-in-Publication Information Available

Library of Congress Cataloging-in-Publication Data Available

ISBN 978-1-63076-131-8 (hardcover)
ISBN 978-1-63076-132-5 (e-book)

∞™ The paper used in this publication meets the minimum requirements of American National Standard for Information Sciences—Permanence of Paper for Printed Library Materials, ANSI/NISO Z39.48-1992.

Printed in the United States of America

This is a story for Gramps
Louis Joseph Tomaselli

CONTENTS

Omnia mirari, etiam tritissima.
[Find wonder in all things, even the most commonplace.]
—Carolus Linnaeus

Introduction

Relax.

It's not actually possible to get poison ivy from opening this book.

It is, however, possible to get poison ivy—the exquisite torture of that insanely itchy rash—in a bewildering host of other ways. The inconspicuous little plant contains one of the most potent toxins on earth. A chemical called urushiol, found throughout the plant, can cause an allergic reaction in approximately 85 percent of humans, and dermatologists have estimated that one ounce of urushiol would be enough to give a rash to thirty million people. The stuff is plant plutonium.

Even if you haven't had the pleasure of meeting poison ivy in person, you've heard tales of it, I'm sure. For centuries poison ivy has bedeviled, inconvenienced, and downright tortured the human race. But *why* is poison ivy apparently on a mission to give us grief, from itchy ankles to life-threatening medical emergencies? What is the purpose of a plant expending precious energy to produce such a powerful toxin?

It isn't hard to imagine that it's all an evil plan—nature's revenge, so to speak. Because poison ivy goes out of its way to *target humans.*

Yes, indeed. Just us. The astounding paradox is that poison ivy is a plant of immense ecological value. Wild mammals from mice to moose, livestock like goats and cattle, hundreds of species of insects—all defy poison ivy's nasty nature and feast on its leaves. Bees and butterflies suck its sweet nectar. (Yes, it has flowers.) Woodpeckers, wild turkeys, robins, and bluebirds all feed with gusto on its fruit. Cardinals are even known to line their nests with fuzzy poison ivy rootlets, in which the young birds nestle comfortably.

Don't all these creatures get itchy?

Nope. Just us.

Over the years, after many bouts of blistered ankles and sore elbows, I've learned to look for poison ivy wherever I step. And, as often follows, with increased awareness comes . . . well, perhaps not love, but . . . let's say, appreciation. I've developed a grudging admiration for poison ivy's beauty, savoring its ruby leaves in spring, its gold and scarlet plumage in fall. I've found I have to watch my step almost anywhere I go: poison ivy is a plant with a broad tolerance for an enormous range of habitats, thriving coast to coast, from mountains to desert. I've seen the familiar three-parted leaves lurking in dark Canadian forests and sprawled in the sun on the beaches of Cape Cod. I've avoided their toxic touch in Appalachian wilderness and in Manhattan's Central Park. Poison ivy hangs out in a humid Louisiana bayou just as comfortably as it does in Arizona desert, where in fall the crimson leaves glow like Christmas tree lights against the canyon walls.

Okay, pretty leaves, yes. But. Still. That itchy problem just won't go away. Once you've tangled with poison ivy, you never quite forget it. I think it's fair to say that no other plant in the world has been detested with such bitterness. It comes as a surprise, therefore, to find that poison ivy was for centuries a popular—and very expensive—garden plant. I'm serious. There was a time, centuries ago, when seeds of poison ivy were almost literally worth their weight in gold. Planted in the gardens of emperors, presidents, and kings, poison ivy was displayed like a captive tiger equally prized for its beauty and its deadliness.

Alas for poison ivy, that's no longer the case. Understandably so—no one wants a plant that causes a horrible rash twining around the picnic tables, the swimming pool, or the playground swings. So poison ivy is drenched with herbicides in backyards and schoolyards, hiking trails and campgrounds across the nation. I would guess that few plants on earth undergo such a barrage of deadly chemicals. Yet poison ivy remains.

What's the secret of poison ivy's success? Thousands of other species of plants are endangered: ferns, wildflowers, trees, vines. It might be worthwhile to examine how it happens that this particular plant can defy humans' best efforts to exterminate it—and it might be worth our while to give a thought to the possible effects of spraying herbicides around the swing sets, the swimming pool, and the picnic tables.

In writing this book, I've followed poison ivy's trail as the despised plant weaves its way through history and over the landscape, and the meandering vine has led me into places I never thought to go. You can't research poison ivy without delving deep into the history of the American wilderness, and the American backyard, and discovering how they've changed over the past three centuries. You can't study poison ivy without learning about the complicated tangle of its interrelationships with the wildlife it nurtures, and with plants that it helps or harms.

Poison ivy, I've found, is a lens through which to take a broader look at the whole green world around us. The human love/hate relationship with this plant is a microcosm of our changing attitudes to nature over the centuries. Poison ivy's ups and downs mirror our drastically changing landscape and climate. And poison ivy's future is rosy: turns out the plant thrives in the exact conditions that climate change is now imposing on our stressed planet. A plant, perhaps, worth studying.

Most of all, I've discovered that the story of poison ivy is inextricably intertwined, for good and ill, with people. It's part of our history, a plant both adored and reviled by our ancestors. We're locked in a toxic battle with it here and now. And in spite of all our efforts to eradicate it, I'm absolutely certain that poison ivy will be there in the future, twining around the ankles of our descendants.

CHAPTER I

The Poysoned Weede

Jamestown, Virginia, 1607

WHEN CAPTAIN JOHN SMITH FIRST SET FOOT ON THE SHORE OF THE New World, he had little idea of what hazards awaited him. "Virginia is a country in North America," Smith wrote of the great green land that stretched before him on that spring morning. "The bounds thereof on the east side are the great ocean. . . . but as for the west thereof, the limits are unknown." Smith and his band of nervous colonists hoped to reach the Pacific Ocean in a few days' journey. They had only the haziest conception of what lay over the horizon—or just around the corner.

John Smith was certain, however, that enemies lurked in this new land. He had read accounts of previous settlers, and he expected to find Virginia thickly populated with "savages." So Smith was wary and well-armed with musket, sword, and metal breastplate. He was prepared to face any dangerous natives, and he knew they could be unpredictable, ubiquitous, and cruel.

But Captain Smith had human adversaries in mind.

⸻

The three-leaved plant, creeping up tree trunks and spreading over the forest floor, appeared harmless at first. Smith, an unusually observant explorer for his time, noticed the inconspicuous leaves that brushed against his boots as he strode into the forest. He commented that it was "much in shape like our English Ivie," the graceful vine that covered the walls of manors and churches back home. However, it wasn't the same old ivy he was used to, as he soon discovered.

Poison ivy (*Toxicodendron radicans*) and its relatives, poison oak and poison sumac, were completely unknown to the Old World. Europe, of course, is well endowed with thorns, thistles, and stinging nettles—plants that let you know in no uncertain terms when you've encountered them. But the colonists were unprepared for a plant that could wreak so much havoc so stealthily.

Poison ivy's chemical warfare against the human race is surprisingly subtle—when poison ivy attacks, you don't realize what's happening. The leaves caress your skin as you walk past, and only hours or days later do the results become clear. A chemical compound called urushiol, found throughout the plant, is a powerful allergen that triggers a delayed dermatitis reaction in most—but mysteriously, not all—humans.

And Smith, like many other overconfident explorers, fell into the plant's trap. "Being touched," he wrote, "the poysoned weed causeth rednesse, itchynge, and lastly blisters." John Smith was certainly not the first human to get poison ivy, but he was the first to describe the plant and its resultant rash in writing, and his description of the symptoms is clinically exact—it's not hard to imagine that he wrote from firsthand experience.

Short and stocky, with a big red beard that almost hid the metal breastplate on his chest, John Smith was a young man when he joined the Jamestown expedition—only twenty-seven. But he already had years of military experience under his belt, and had travelled thousands of miles as a mercenary, adventuring in exotic places like Transylvania, Russia, and Constantinople. The thought of exploring a wilderness, even one laced with poison ivy, didn't faze him at all.

He and the other would-be colonists, a group of just over a hundred men and boys, were bankrolled by the wealthy merchants of the Virginia Company of London, who were hoping to make a financial killing on the venture. The colonists were also eager for quick riches—in fact, everyone involved in the venture had, as Smith irritably put it, "great guilded hopes." In the forests of Virginia, green as the Garden of Eden, many Englishmen assumed that gold would be theirs for the taking, handed over by friendly natives eager to gain the benefits of civilization. But the

Scratching the Itch: Old World Remedies

"*Rednesse, itchynge, and lastly blisters . . .*" If any of the James-town settlers experienced the unpleasant dermatitis reaction, what might they use to cure it?

Any of the better-educated colonists (including Smith, a very well-read man) would have been familiar with a best seller of the time: *The Herball, or Generall Historie of Plantes*, a weighty 1,480-page volume of botanical cures. Writing in 1597, John Gerard wasn't yet aware of poison ivy, but he listed dozens of remedies for itchiness and skin conditions. However, all of his suggestions called for the use of Old World plants which at first weren't available to the itchy colonists of Jamestown.

But that was soon to change. One sovereign remedy that Gerard recommended for rashes and bug bites was an English roadside weed known as plantain (*Plantago major*). Plantain was one of the very first plants transported from Europe to America, probably brought for herb gardens because of its well-known medicinal qualities. The plant appeared in America shortly after the Jamestown colonists' arrival and was well-established by the time of the first botanical survey of the New World, in 1671.

Gerard wrote that plantain leaves "are singular goode to make a water to wash a sore place, or the privy parts of a man or woman." The plant was much used by colonists and soon adopted by Indians as a quick and easy poison ivy soother.

Plantain still works today, and you can find testimonials to its efficacy all over the Internet. Making an infusion of the leaves steeped in hot water, as Gerard recommends, can be beneficial, but the plant is an effective trailside remedy, too. Simply crush-ing a few of the leaves and placing them on the skin can help to ease insect bites or itchy rashes. (See appendix for much more on poison ivy remedies.)

colonists should have remembered that in Eden, things were not always what they seemed.

Smith viewed the new terrain with a slightly different eye than the gold-seekers. Although certainly not averse to getting rich, he was a thoughtful adventurer who realized that you had to check out any potential Garden of Eden for possible serpents. Not even poison ivy could dim his enthusiasm for exploration—after all, as he gamely pointed out, the effects of poison ivy "after a while passe away of themselves without further harme."

Most of the Jamestown settlers were terrified of encountering the dreaded "savages." They promptly built a sturdy wooden palisade for defense, and rarely ventured outside its walls except to prospect for gold. But John Smith eagerly headed into the unknown, not to search for treasure, but to investigate the forest that surrounded the tiny new settlement. He was about to walk into paradise, indeed.

The Old World that Smith had left behind was an environment highly altered by humans. The dense forests of the Bronze Age had long been felled, and much of Europe had been heavily farmed for centuries. Rivers that flowed near human habitations were generally little better than streams of sewage. Sprawling and ever-growing cities were interspersed with overcultivated, thin-soiled farmland. To Smith and the wide-eyed colonists, the woodlands of the New World were a breathtaking sight.

"Heaven and earth never agreed better to frame a place for man's habitation," Smith wrote. "Here are mountaines, hills, planes, valleys, rivers, and brookes all running most pleasantly into a faire Bay, compassed . . . with fruitfull and delightsome land." Interspersed with meadows rich with wildflowers and berries were groves of trees tall as church spires. Smith immediately recognized familiar species, oaks and pines and so forth. But these oaks were ten feet in diameter, and the centuries-old pine trees seemed to touch the sky. Then there were the *big* trees.

Awestruck, Smith wrote of "Cypress trees 80 feet high without a branch . . . oakes so tall and straight, that they will beare two foot and a halfe square of good timber for 20 yards long . . . chestnuts whose wild fruit equalize the best in France, Spaine, Germany, or Italy." There were wild cherries and plums "as delicious as an Apricock." Many mouthwatering species were unknown to the Old World: sweetgum, sassafras, butternut, black walnut, persimmons, and pawpaws.

And there were flowers: massive magnolia trees with creamy blossoms the size of dinner plates; flowering dogwood, silverbells, redbud. On the forest floor shone constellations of bloom: orchids, wood lilies, bluebells scattered among ferns and moss. And these botanical splendors were well festooned with impressive poison ivy vines, which, if left to grow undisturbed, can have a circumference of twenty inches and be a hundred feet in length.

A hundred feet of poison ivy seems like a terrible thing. But from the point of view of the wildlife that abounded in the New World, poison ivy was—and is—a blessing, not a curse. From the tiny wren to the portly wild turkey, birds flock to feed on poison ivy's white, waxy berries. In fact, poison ivy attracts a rainbow of avian life: purple finches, bluebirds, ruby-crowned kinglets, yellow-shafted flickers. A dozen species of warblers use the berries to fuel their long migratory flights. Poison ivy leaves are food for a host of wildlife: black bears, rabbits, hordes of insects. White-tailed deer actively seek it out for browse, preferring it to many other plants.

Scratching the Itch: They *Eat* Poison Ivy?

While the idea of eating poison ivy may make us humans gag, no other species is affected by poison ivy as drastically as we are. Humans' intense allergic reaction to urushiol isn't shared by our nonhuman neighbors. The process of chewing and swallowing releases abundant amounts of the itch-causing chemical urushiol, but poison ivy leaves and berries vanish into the beaks and mouths of countless species with no apparent ill effects.

When Smith set out on his first venture into the uncharted forests of Virginia, he assumed that the land he was exploring must be "a plaine wildernes as God first made it," all but untouched by human hands. But the landscape he travelled through had been shaped and altered extensively by its human inhabitants: the many Native American tribes scattered thickly along Virginia's coastal region. An estimated twenty-five thousand Algonquin-speaking people were living in the Virginia coastal region, calling their realm Tsenacommacah, which means "densely inhabited area." And their activities had created an excellent habitat for the abundant growth of poison ivy.

Smith was enchanted by the beauty of the place he was exploring, but pointed out that the country would be ideal "were it fully manured [farmed] and inhabited by industrious people." To his mind, the Indians weren't "industrious" enough to get out there and dig and fertilize and make use of the land properly. He saw no barns, no fences, no hardworking farmers sweating behind ox-drawn plows. What Smith didn't understand was that the Indians were indeed altering the land in their own way. Instead of metal tools and domestic beasts of burden, Indians used nature's own force to shape the environment—the power of fire.

Only two days after sighting the coast of Virginia, the colonists had observed "great smokes of fire" rising from the trees. "We marched to these smokes," wrote George Percy, one of the colonists, "and found that the savages had been there burning down the grass." The jumpy newcomers mistook the burning for smoke signals presaging an attack, but in fact the Indians were engaging in their own particular sort of agriculture. The Virginia Indians burned thousands of acres every year, resulting in a very different landscape than the virgin forest of prehuman times.

Smith frequently observed wide glades free of underbrush, where he noted that a man could gallop a horse unobstructed: conditions created by frequent, low-intensity ground fires. The Indians had learned how to set blazes that burned quickly, incinerating weeds and brush while doing minimal damage to large trees. These fire-cleared meadows grew into rich pastures for deer to graze, as well as open land easier to work for crops.

Poison ivy thrives in meadows bordering woodlots, with sunlight to encourage germination as well as trees for the weak-stemmed vine to climb. And poison ivy is especially well-adapted to fire. Although poison ivy's leaves and thin-barked stems are easily killed by burning, its vast network of underground roots survive and resprout with enthusiasm. This meant that there were even more poison ivy fruits and leaves available for birds and other wildlife than in an untouched, primeval forest.

One clue that poison ivy was a very common plant edging the burned-over crop fields, berry bushes, and hunting grounds is the large number and remarkable variety of traditional Native American remedies for poison ivy. Apparently John Smith wasn't the only one dealing with "rednesse, itchynge, and lastly blisters."

Scratching the Itch: New World Remedies

Indians were undoubtedly far less likely than Englishmen to go blundering into a patch of poison ivy—but anyone can make a mistake. Fortunately, there was relief to be had in the rich botanical pharmacy growing all around. Ethnobotanical records of many tribes list a host of remedies for poison ivy.

For relief, each healer turned to the plants that grew in his or her own particular neck of the woods. Many groups of forest dwellers in the Northeast used jewelweed, a common deep-woods plant. Jewelweed's dangling orange flowers grow on stems with a clear watery juice that soothes irritated skin. Lenape living along the Delaware River used an infusion of sweet fern that grew along streams and riverbanks. Iroquois in the foothills of the Adirondacks often turned to the healing sap of white pines. The Mohegans, and other agricultural tribes as well, made an infusion of dried corn cobs as a soothing wash for a poison ivy rash. (For more details on the efficacy of these and other herbal remedies, see appendix.)

The Jamestown settlers scratched their rashes, pitched tents and built huts, and prospected mightily, not realizing at first how closely their activities were being watched. There wasn't a lot of unpopulated land to spare, and tensions between the Indians and the Englishmen began from the day the colonists stepped ashore, when in a brief and nasty skirmish, Smith wrote, they "were assalted with certaine Indians, which charged them within Pistoll shot."

But the Indians could be welcoming, too. During Smith's first exploratory jaunt, he visited different villages and slowly began to unravel the Indians' complicated pattern of alliances, friendships, and enmities, as tangled and confusing as any in Europe. There wasn't just one tribe, there were dozens, many of them united in a huge and powerful confederacy. Smith and his companions eventually sailed "up the River to Powhatan, of which place their great Emperor taketh his name, where he that they honored for King used us kindely."

This great chief had ruled his empire for decades; now in his sixties or seventies, Powhatan was a patriarch with dozens of wives and many children. Among them was a favorite daughter named Matoaka, better known by her nickname, which has been translated as "the playful one," "laughing and joyous one," or even "little hellion": Pocahontas. This curious and precocious child was eventually to become famous, her story forever linked with that of Jamestown and John Smith.

Smith, for his time, was surprisingly open to learning from the Indians, and he was especially intrigued by Pocahontas. The little girl (he estimated her to be about ten or twelve years old) came often to visit the cluster of tents and ramshackle huts that the colonists had surrounded with a tall log palisade. Smith even attempted to master her complex language, creating the first Algonquin dictionary, with more than five hundred words and phrases. In Algonquin just one word is needed to describe a situation both colonists and Indians frequently found themselves vexed with: *mitashkishin*, which means "to be infected with poison ivy."

Powhatan women were responsible for gathering medicinal and edible plants, and as a Powhatan girl, Pocahontas would have been learning this essential skill. It's tempting to imagine that she was the one who pointed out the "poysoned weede" and perhaps shared jewelweed or other poison ivy remedies with John Smith.

— · —

After only a few months in America, the eager prospectors of Jamestown were bitterly disillusioned. Virginia, it turned out, was far from being a paradise after all. They had failed to unearth so much as a speck of gold, and the New World seemed to offer nothing but mosquitoes, heat, and poison ivy. Soon food supplies ran low and disease became rampant. By the end of the first year, two-thirds of Smith's shipmates were dead.

But John Smith never lost his enthusiasm for the "fruitfull and delightsome land." After a serious injury from an accidental gunpowder explosion, he returned to England to recuperate, but in spite of—or perhaps because of—its dangers and challenges, Virginia was still on his mind. "I call them my children," he wrote long afterward of the American settlements that haunted his imagination, "for they have been my wife, my hawks, my hounds, my cards, my dice and in totall, my best content."

Smith never did manage to return to Virginia, but he succeeded in bringing the savage beauties of America alive for others. A prolific author, one of his best sellers was an illustrated volume he wrote in 1612 titled *A Map of Virginia with a Description of the Countrey, the Commodities, People, Government and Religion.* He was attempting to attract potential investors in new colonies, so he was writing at least partly as salesman, downplaying the dangers and privations and focusing on the positive. He didn't promise gold mines, evasively saying that "Concerning the entrailes of the earth little can be saide for certainty," and pointing out acidly that with better-trained prospectors it might be possible to unearth some valuable minerals. But he wasted few sentences on "guilded hopes," instead rhapsodizing for pages about the "many excellent vegitables and living Creatures," as well as the fertility of the soil, abundance of fish and shellfish, and the many navigable streams. He emphasized the abundance of edible and medicinally useful plants, enumerating their many benefits. He even had a good word to say for poison ivy. "Yet because for the time they [the rash and blisters] are somewhat painfull, it hath got it selfe an ill name," he admitted. But, determined to look on the bright side, he went on to insist that poison ivy was "questionlesse of no ill nature."

Many of Smith's fellow adventurers had lost their lives in the search for nonexistent gold. John Smith was among the first Europeans to suggest that the true potential riches of North America were not to be found underground, but rather springing from the fruitful earth. Smith knew that many Virginian plants could turn out to have tremendous value when exported to Europe. What he did not foresee was that "the poysoned weede" would soon be heading to Europe, too.

Seven years after Smith abruptly left America, Pocahontas also took a ship across the Atlantic. One of the later Jamestown arrivals, a tobacco entrepreneur named John Rolfe, had become enraptured by her. The pious Rolfe was amazed at himself, questioning with "feare and trembling" the impulse "which thus should provoke me to be in love with one whose education hath bin rude, her manners barbarous, her generation accursed, and so discrepant in all nurtriture frome my selfe." But he could not resist her "to whom my hartie and best thoughts are, and have a long time bin so intangled, and inthralled in so intricate a laborinth, that I was even awearied to unwinde my selfe thereout."

As a way to raise interest in colonization and promote the increasingly lucrative tobacco trade, Virginia Company investors arranged in 1616 for Mr. and Mrs. Rolfe to travel to England. Formally attired in silks and velvets, "Rebecca Rolfe" took London by storm. The rich and titled flocked to meet her, she attended balls and theaters, and she dazzled the leaders of society, including King James I.

About the same time as Pocahontas was hobnobbing with the wealthy and famous, another beautiful American was being transplanted overseas to meet with a similarly enthusiastic reception. Like John Smith and the hapless John Rolfe, many Europeans found themselves "intangled and inthralled" by the wild and strange creatures of the New World—and its remarkable flora. Botanists and doctors, scientists, gardeners, even the kings and queens of Europe were about to discover a new and fascinating plant.

Poison ivy did not migrate to the Old World as a stowaway, seeds accidentally attached to a bootheel or hidden in soil used for ballast. *Toxicodendron radicans* was an honored passenger, so to speak, cordially invited on board. Carefully harvested, nurtured, and transported across the North Atlantic—perhaps on the same ship that carried Pocahontas— poison ivy was about to take London by storm.

CHAPTER 2

A Collection of Rarities

London, England, 1634

THE ENTRANCE TO THE MUSEUM WAS FLANKED BY TWO GIGANTIC whale ribs. Customers passing under this impressive gateway were satisfied that their admission price had been well spent—the wonders they were about to see were certainly worth the expense. The entry fee of sixpence was considered quite pricey: a day's wages, in fact, for many of the attendees. But there was no doubt about it—John Tradescant's museum was a sellout hit.

Londoners flocked to admire the exhibits this intrepid adventurer and his son had gathered from all over the world. Crowds lined up to gawk at the dried mermaid's hand, the unicorn's horn, even an authentic piece of the True Cross. And the viewers were especially fascinated by items from the New World, including Tradescant's magnificent collection of American plants. The Tradescants, father and son, were expert gardeners and they had succeeded in growing in the gravelly London soil a remarkable array of New World flora: bright plumes of goldenrod, fragrant blossoms of black locust, sturdy white cedars, delicate trout lilies, Virginia wild rose—and the beautiful but savage plant called poison ivy.

~~~

John Tradescant is the first person known to have cultivated *Toxicodendron radicans*. A portrait of this famous seventeenth-century English gardener shows a tight-lipped elderly man gazing at us with stern eyes. He wears the plain clothing of a humble man, black with a simple white collar, but the glint of a gold earring in his left ear hints at the spirit of a buccaneer. Tradescant was indeed an odd mix of pirate and gardener, an adventurer who turned his fascination with the colors, shapes, and scents of the green world into a successful career.

The profession of gardener is a humble one—unless you're hired by the rich and famous. As a young man, John Tradescant managed to use good family connections to land the plum job of gardener to the powerful Robert Cecil, Queen Elizabeth's secretary of state and chief spymaster. After Cecil's death, Tradescant was hired by the elegant George Villiers, Duke of Buckingham, who was famed for his fabulous mansions and gardens (Buckingham Palace was just *one* of his residences). After Buckingham's shocking assassination, Tradescant was once more out of a job, but he must have made some more good connections, for he soon moved even higher up the social ladder when King Charles I appointed him Keeper of His Majesty's Gardens, Vines, and Silkworms at Oatlands Palace in Surrey.

During his career as gardener, Tradescant didn't pull weeds and trim hedges. He oversaw the creation of magnificent estates. And his celebrity employers wanted to have the newest, rarest plants for their showplaces; they demanded ever more exotic plants from Asia and Africa as well as the two great, mysterious continents that were just beginning to be explored—the Americas.

All the maps, it had turned out, were wrong. For centuries Europeans had trusted the ancient Greek and Roman cartographers, who had of course been totally ignorant of the existence of land to the west between Europe and Asia. Then the discoveries of Renaissance explorers turned accepted patterns of thinking upside down and contradicted ancient sources of wisdom. Suddenly whole new continents existed beyond the horizon, and according to tantalizing hints that were coming back as John Smith and others published tales of their adventures, these new lands were filled with strange and wonderful things. Fierce beasts! Savage warriors! Amazing plants! But only a tiny percentage of seventeenth-century folk would ever make the long and dangerous journey to see this fabulous New World. People were hungry for a glimpse of marvels. John Tradescant scented a business opportunity.

From the start, Tradescant was especially fascinated by America's rich flora. He was one of the early investors in the Virginia Company, and a close friend of John Smith. Tradescant doubtless heard many stories of the entrancing plants of the Virginia wilderness from the famous adventurer.

Tradescant never saw poison ivy in its natural habitat, since he never made it to America himself, but he was no stay-at-home. All his long life, the white-bearded gardener had a childlike sense of wonder and a fascination with "anything that is strange," and on his expeditions, he always had an eye out for rarities, be they animal, vegetable, or mineral. He roamed the world collecting oddities like a child picking up bright pebbles on a beach.

He went on missions to collect plants for his noble employers all over Europe, even venturing as far as Siberia, from whence he brought back the sweetly scented Muscovy rose. He also went to war; during an expedition against Barbary pirates, he participated in sea battles, but the highlight of his trip was collecting specimens of a new sort of sweet apricot. He accompanied the Duke of Buckingham to war in France, and although the campaign was disastrous, Tradescant was delighted to discover a new and odd variety of prickly strawberry.

Tradescant also networked, using his contacts with nobility to gather specimens for his growing collection. In his privileged position as the Duke of Buckingham's gardener, he penned a letter to the secretary of the navy, petitioning aid from "Marchants from All Places, But Espetially the Virginie & Bermewde & Newfound Land." Tradescant requested that merchants travelling to America be asked to seek out "All Maner of Beasts & Fowels & Birds Alyve . . . Heads Horns Beaks Claws Skins fethers . . . flyes or seeds plants trees or shrubs . . . that be Rare or Not known to us." Specimens began to trickle in, and Tradescant's collection of marvels was growing.

In the 1620s he and his family moved into a big house in Lambeth, a rural suburb of London, and Tradescant began creating a garden of his own. He was now approaching sixty, and his voyaging was done: it was time to organize his collection of plants and begin his garden.

The soil in the area was thin and pebbly, so many of the local market gardeners dug up gravel and sold it for repairing roads or ships' ballast, then replaced the gravel with "the filth of the city, which serves for excellent manure, rich and black as thick ink." More and more plants swelled Tradescant's collection in this smelly but fertile soil: trees, ferns, flowering shrubs, vines. In 1634 he took an inventory of his garden, which he published in a pamphlet titled *Plantarum in Horto Iohannem Tradescanti nasccentium Catalogus.* It listed 770 species and varieties grown at Lambeth, including forty Virginia species, among them poison ivy.

Unfortunately, he failed to list the sources of his collection, so we'll never know who the brave or foolhardy person was who first dug up poison ivy rootlets or extracted seeds from the tiny white berries and brought them to the Old World. It could have been one of his friends, John Smith or Samuel Argall (captain of the ship that brought Pocahontas to England). It may even have been his son, John the Younger, although the date of the younger Tradescant's visit to America is uncertain, and many of the species he brought back went unrecorded. At any rate, poison ivy was one of Europe's first New World acquisitions. And John Tradescant saw the irritating vine as a prize, not a curse; one of nature's rarities, an oddity well worth collecting.

Many wealthy noblemen collected nature's rarities. But up till then, the thrill of seeing unicorn horns, hippopotamus teeth, and exotic orchids had been reserved for the well-born. John Tradescant broke with this tradition: he invited one and all to come to Lambeth—as long as they paid the admission fee. "Tradescant's Ark," as the institution was affectionately called, was the first museum in Europe that was open to the public—perhaps the first in the world. Nobles, housewives, clerics, professors, rich ladies, merchants, apprentices out for a spree—everyone came to see the sights.

Lambeth is now part of London's urban sprawl, but in Tradescant's day it was a green and pleasant place across the Thames River. Patrons could walk across London Bridge and escape the city's clamor, strolling past farms and marshlands dotted with windmills—or if wealthier, they could pay the fare to take the horse-ferry, or hire a comfortably cushioned

Scratching the Itch: Gardening with Poison Ivy?

He grew poison ivy? Poison ivy—the plant that can give you a painful rash with blisters, weeping sores, and an itch that lasts for weeks. Was John Tradescant insane?

Tradescant left no record of his planting methods, so we'll never know exactly how he dealt with poison ivy. Was he immune? Careful? Or did he merely assign an assistant gardener to do the dirty work for him?

Believe it or not, there are several reasons why poison ivy is, to some extent, gardenable. First, not everyone is equally sensitive to poison ivy. About 15 percent of the human population is, as doctors put it, "exquisitely" allergic. The slightest contact with poison ivy can give these people a severe dermatitis reaction. And a small percentage—again, about 15 percent—are immune, and a gardener who chanced to fall into this lucky category could handle it with impunity. For myself, I'm one of the 70 percent in the middle, who get a mild rash but don't have tremendously unpleasant symptoms. Speaking only for myself, I have found that I can handle poison ivy if I'm very careful about not letting any part of it touch my skin. I wear gloves, which I throw away— don't even attempt to wash them. Not that I'm recommending that you go out and add poison ivy to your flower bed, but for some people, it is possible to handle the plant.

Once rooted, it's quite easy to grow, very low maintenance indeed. Poison ivy is a long-lived perennial, and it doesn't require all that tiresome pruning, staking, dividing of roots, or what-have-you that so many ornamentals need. Once the poison ivy was draped artistically across a trellis, there wasn't much John Tradescant had to do to maintain it.

Tradescant was an expert gardener, a professional at the peak of his long career and well used to handling odd, possibly dangerous exotics filled with poisons or covered with thorns or spikes. So it's not impossible to imagine how he could have coped with poison ivy.

Or then again, perhaps he was one of the fortunate 15 percent.

boat to float them downriver. After passing through the whale-rib gate and paying their sixpence, visitors entered the huge garden.

Tradescant loved fruit trees, and was famous for his delicious specimens of apples, plums, and apricots. He had many magnificent trees from the New World as well. In autumn the North American plants were especially spectacular, as the three-parted leaves of poison ivy twined among bright goldenrod, royal purple asters, and the deep red fruit clusters of staghorn sumac.

Georg Christoph Stirn was a German tourist who in 1638 was among the many visitors to the Ark. He wrote a vivid account of his sightseeing, describing "all kinds of foreign plants" in the garden. He also enjoyed the other rarities, which were kept in a big room upstairs. Stirn wrote of exotic creatures (probably stuffed, not alive) such as "a salamander, a chameleon, a pelican . . . a flying squirrel, another squirrel like a fish . . . a sea parrot, a toad-fish . . . a bat as large as a pigeon." He also admired "a small piece of wood from the cross of Christ" and "the passion of Christ carved very daintily on a plumstone." No wonder the Ark was deemed one of the main tourist "must-see" spots of London for decades.

Poison ivy was not only one of the very first New World plants Tradescant collected, it was apparently one of his and his son's favorites—at any rate, it was maintained on the inventory lists of the museum for years, as the younger Tradescant carried on the family business long after his father's death. Perhaps they realized that to London museumgoers the savagery of the New World was one of its main charms. Poison ivy's deadly reputation made it more alluring than ever—part of a biological Chamber of Horrors, along with other titillating and gruesome specimens from the New World, which Stirn described with relish: "a piece of human flesh on a bone . . . a human bone weighing 42 lbs. . . . Indian arrows such as are used by the executioners in the West Indies—when a man is condemned to death, they lay open his back with them and he dies of it . . ." The public loved it.

The Ark was acquired after the passing of both Tradescants by a rival collector named Elias Ashmole. The rarities were taken from the Lambeth house by the avaricious Ashmole, who reverted to the custom of showing them only to the aristocracy. As the years passed, the Tradescants' garden of North American plants was forgotten. An inventory taken years later noted staghorn sumac still present among the weeds, as well as a huge cypress, which endured for more than a century. Poison ivy was not mentioned.

Once hailed as a "great treasurer of nature's rarities," John Tradescant was also all but forgotten. But his collection formed the nucleus of what is now one of London's premier museums, the Ashmolean. Many of the Ark's curiosities are still enshrined there, including a preserved dodo skin. But if John Tradescant himself gradually faded into obscurity, poison ivy's fame was just getting started. And soon the rare plant would be a very valuable commodity.

CHAPTER 3

Bartram's Boxes

Philadelphia, Pennsylvania, 1746

THE BIG WOODEN BOX WAS PACKED TO THE BRIM, ALMOST READY FOR shipping. A neatly dressed Quaker gentleman, spectacles balanced on his nose, surveyed the contents of the crate with satisfaction. The box was filled with seeds—thousands of seeds.

He had made an inventory of the contents, noting each species carefully in his erratic spelling. Now he checked the list over, to be sure that the customers who were eagerly awaiting this box would get full satisfaction.

Yes, everything was in order. There were bags of acorns, boxes of the papery winged seeds of maples, heaps of spicy-smelling cones that would yield firs and spruces, cedars and pines. Some seeds were folded in paper, some wrapped in moss or embedded in jars of sand. He carefully added a packet filled with tiny seeds of poison ivy.

Nailing the lid down tightly, John Bartram nodded with satisfaction. Another five guineas to put in the ledger, another shipment embarked across the sea. Another one of Bartram's boxes was on its way.

— ~ —

Throughout John Bartram's long and adventurous life he was fascinated by plants. Even as a boy, he frequently played hooky from the family farm, and would run off to the woods and revel in ferns, mosses, and wild-flowers. As he grew up, Bartram appeared to be just another industrious

Quaker farmer. He plowed and harvested his fields, attended worship services, eventually married and raised nine children. But he still had that odd habit of occasionally wandering off to the woods and collecting wild seeds to grow in his garden.

Bartram networked with many other nature enthusiasts of the time, including Benjamin Franklin, sending letters describing his hobby as well as samples of seeds. So when a British gardener named Peter Collinson was looking for someone to send him a collection of American plants for an experimental garden, John Bartram was the obvious choice.

At first Bartram shipped to England whatever he picked up on his ramblings around Pennsylvania—an unlabeled jumble of seeds. Collinson had to plant the mystery seeds and wait to see what happened. But he was thrilled by these new, exciting plants and beseeched his American supplier to send more—and to get the system better organized.

The Englishman sent over thick stacks of herbarium paper (a very expensive commodity), urging Bartram to mount and label a specimen of each plant he obtained seeds from, so that whoever planted the seed would know what might emerge. Collinson also suggested shipping live plant roots wrapped in watertight ox bladders, the eighteenth-century version of a plastic bag.

Bartram built sturdy rectangular shipping boxes, about three feet long and two feet high, nailed together from rough boards. Some of the boxes had padded leather collars to hold young plants upright and protect tender stems and fragile shoots. He had special compartments for seeds, and tiny drawers for roots and bulbs.

We don't know in what form he first shipped poison ivy across the Atlantic. If he sent only the seeds, it's likely Collinson would have been frustrated. Female poison ivy plants produce clusters of single-seeded fruits, known botanically as *drupes*, which look like small berries. The hard little seeds inside the fruit are *indehiscent*, which means they don't automatically pop open on their own when planted, as most seeds do. Poison ivy seeds can't germinate until they're scarified in some way, charred by fire or roughed up by the process of passing through the digestive systems of birds. Just sticking them in potting soil wouldn't do the trick.

Bartram probably shipped poison ivy rhizomes as well, which are underground stems that shoot off from the main stem. The fuzzy brown rhizomes spread rapidly just below the surface of the soil to form new plants, and would undoubtedly be the easiest way to propagate a nice, healthy crop of poison ivy.

No matter how carefully he packed them, Bartram's sturdy boxes didn't always get to their destination. Shipping plants is always a chancy business, as the good folks at the post office know only too well; shipping via wooden sailing ship, the odds against plants arriving intact were high. Sometimes the ship was lost in a storm, and the seeds ended up at the bottom of the ocean. Some vessels were captured by pirates, who were uninterested in botany and heaved the precious boxes overboard. One shipment of seeds, hopefully not including poison ivy berries, was opened and eaten by hungry cabin boys. Even if the ship reached port successfully, some of the plants and seeds often turned out to have been destroyed by salt water or mold, or nibbled by rats. But Bartram, undeterred, kept on collecting and sending plants.

Collinson graciously shared the goods with other plant enthusiasts. Soon word began to spread. Naturalists throughout Europe eagerly awaited the arrival of these new and exciting specimens. All through the scientific world, the big wooden cases packed full of botanical marvels became known as Bartram's boxes.

At first Bartram's customers were scientists like Carolus Linnaeus and Sir Joseph Banks, erudite scholars who were studying and classifying flora. But European interest in American plants was growing like a weed. Wealthy landowners were willing to pay big money for impressive plants like American oaks and pines, sugar maples and elms, gorgeous flowering shrubs like mountain laurel and rhododendron, and—believe it or not—poison ivy.

It wasn't that people didn't know that gardening with the plant could be problematic. In 1668 Bermuda colonist Richard Stafford sent some poison ivy to a collector friend in England, and he took care to explain the unpleasant effects clearly: "I have seen a Man, who was so poyson'd with it, that the skin peel'd off his face, and yet the Man never touch'd it, onely look'd on it as he pass'd by." Yet in spite of these warnings, poison ivy's reputation as a remarkable plant began to spread through Europe.

Increasingly, American plants were all the rage. At first European gardeners had vied to see who could grow the first pineapples, hibiscus, or coconut palms. During cold snaps, gardeners' apprentices bundled palm trees in straw overcoats and stayed up all night stoking fires to keep orchids and banana trees from freezing. But only the very wealthy could afford the enormous glassed-in conservatories required for tropical plants. North American specimens didn't need such babying; plants that hailed from the mountains of Virginia or the rocky soil of New England weren't bothered by the comparatively mild weather of the Old World. These sturdy Americans could escape from the glassed-in walls of the greenhouse and survive outdoors, even in winter, anywhere on the estate. Poison ivy, a species that tolerates a wide range of habitats, was particularly easy to grow, flourishing in English pleasure gardens as luxuriantly as it had in Virginia wilderness. Poison ivy sank its roots deep in Old World soil.

Now, perhaps you're thinking that poison ivy seems an unlikely candidate for a showplace garden. While the plant is quite pretty in early spring, its tiny new leaves reddish and shiny, by summer the foliage turns to a nondescript green, blending invisibly with the rest of the shrubbery. It's a very inconspicuous plant, really, often spreading at ankle height as a low ground cover as well as climbing trees. Poison ivy is in fact so unremarkable for most of the year that few people recognize it, even as it caresses their ankles or rubs against their elbows.

But as the days shorten and the chill begins to set in, poison ivy undergoes a startling change. While most plants are still clad in the faded hues of late summer, poison ivy leaves turn a rainbow of scarlet, purple, and yellow—one of the earliest plants of the season to put on bright autumn dress.

This botanical rainbow began to yield—unlike most rainbows—a satisfying pot of gold. John Bartram got more organized in his overseas shipping and eventually standardized the "five-guinea box," which had an inventory of more than a hundred species of trees, shrubs, and vines, including poison ivy. Each year he shipped dozens of boxes, sent out like clockwork every fall, to arrive in Europe at the cusp of the planting season in earliest spring.

His British friend Peter Collinson was a highly sociable fellow, a frequent guest of wealthy landowners, and a very efficient networker. He arranged for more and more customers to purchase Bartram's boxes, until the client list included the Dukes of Cumberland, Marlborough, Norfolk, and Richmond, nine different earls, the Prince of Wales, and King George III. Bartram also supplied the elite nurserymen of London, who then exported stock to Sweden, Germany, Scotland, and France. Some impressive Americans patronized his garden, too, including Benjamin Franklin, George Washington, and Thomas Jefferson.

Among Bartram's most valuable wares were the "weeds," common as mud, that you can see in vacant lots and back roads today, especially ones with bright fall colors: New England asters, goldenrod, and staghorn sumac, as well as poison ivy. Bartram began to cultivate these plants in his backyard garden, so that he could easily have stocks at hand.

The demand for Bartram's boxes continued to grow. Some great landowners even strove to re-create whole American forests, poison ivy and all. An English nobleman named Lord Petre, an enthusiastic admirer of American plants, ordered seeds from Bartram on an enormous scale—red cedar berries by the bushel, wagonloads of pine cones. On the nobleman's estates, an army of gardeners planted more than forty thousand American trees, shrubs, and vines, the vast majority of the seeds supplied by John Bartram.

To keep up with the demand, Bartram spent weeks each year travelling through the wilds collecting plants. His insatiable customers kept demanding new species until Bartram wrote snappishly to Collinson, "Do they think I can make new ones? I have sent them seeds of almost every tree and shrub from Nova Scotia to Carolina."

Bartram had many mouths to feed—raising nine children can be expensive—and the money brought in by his seed shipments was surely welcome. But money alone can't explain the enthusiasm with which he collected plants. "The Botanick fire set me in such a flame as is not to be quenched untill death," wrote this quiet and mild-mannered Quaker. He travelled hundreds of miles on horseback into unmapped wilderness, the domain of wolves and bears. Almost always travelling alone, he climbed mountains, waded rivers, and explored the roadless territories of Indian tribes.

Bartram often referred to himself as a fearful man—he was particularly scornful of his own dread of lightning. But in spite of its dangers, the wilds lured him still. Even as he aged, he continued his travels, wandering all along the eastern seaboard, still drawn to the wilderness as though to a lover.

The primly dressed Quaker was particularly adept, all his life, at climbing trees. This was an indispensable skill for a seed collector, as many seeds ripen attached to high branches and vines, waiting for birds or squirrels to eat and so disperse them—as with poison ivy fruits, which are often found among the highest branches. Bartram needed seeds in such quantities that he couldn't just collect the few handfuls that fell to the forest floor, but he rarely resorted to the vandalism of cutting down a tree to get its fruit. At the age of sixty-two, he scaled a huge old holly tree to gather the scarlet berries (leaving his twelve-year-old son at the bottom, holding a collecting bag). When a branch broke, Bartram fell out of the tree and injured himself severely, probably fracturing several ribs. But nothing could discourage his quest for unusual plants.

He especially loved nature's oddities. As well as gathering bushels of pine cones and acorns, he spent much time and energy ferreting out the unusual. The carnivorous Venus flytrap plant was one of his most popular finds. He was fascinated by rattlesnakes and deplored the habit of killing them whenever they were found. On one occasion, while travelling with a guide, Bartram pleaded (in vain) for the life of a particularly large and beautiful rattler they had encountered, even though he had narrowly escaped being bitten.

To John Bartram, poison ivy was not unlike a rattlesnake—something to be respected and approached with caution, yes, but just another example of nature's marvels, more kindling for the "Botanick fire" that burned within him. In an era when most people saw the wilderness as a detriment, an obstacle to be removed, Bartram found endless beauty and wonder in nature. Perhaps he was thinking of his encounters with poison ivy when he wrote: "Whatsoever whether great or small, ugly or handsom, sweet or stinking, everything in the universe in thair own nature appears beautiful to mee."

CHAPTER 4

Loathesome Harlotry

Amsterdam, Netherlands, 1736

THE THREE DRIED LEAFLETS, CAREFULLY PRESSED, SHOWED UP CRISPLY on the smooth white paper. Karl Linné, an eager young botany student who had just landed his first job, was cataloging the plants in his employer's garden, pasting each specimen down flat on a stiff herbarium sheet. He spread the three-parted poison ivy leaves carefully across the page, pondering their shape and color, considering the structure of the plant. He finally decided on a new name to bestow on this plant: *Rhus radicans.*

Carolus Linnaeus—to use the Latin form of his name—had come one step further on his self-appointed mission: to name every living thing in the world.

The great Swedish naturalist was enthralled by plants, almost from birth. Even when he was an infant in his cradle, his parents could soothe him by showing him a flower. His father, an avid gardener, taught his son the names of all the blossoms in his garden, and the boy never outgrew his passionate love affair with "Flora." He blended his study of botany with a medical degree, using his beloved plants as medicines.

But botany was made confusing, and indeed often infuriating, by the bewildering number of names that could be borne by a single plant.

Scratching the Itch:
Ways to Say "Poison Ivy"

By Linnaeus's time, poison ivy had already gotten itself a bad name in many languages and cultures. A Spanish conquistador, who had undoubtedly experienced the plant firsthand, cursed it with the name of *hiedra maligna*, or "malignant ivy." A seventeenth-century Dutch physician, who was less vengeful, or possibly immune to urushiol, christened the plant with the pleasant name of *edera trifolia Canadensis*, or "three-leaved Canadian ivy." John Tradescant recorded it in his inventories as *Frutex Canadencis Epimedium folio*. In Mexico poison ivy was (and still is) known as *mala mujer*, "the wicked woman."

Many plants had a dozen common names, varying widely from region to region. And new names were constantly springing up, as every would-be botanist merrily created impressive Latin or Greek names. Some plants had names that were nine or ten words long. The humble tomato, for instance, a newly imported American plant, was dubbed *Solanum caule inermi herbaceo, foliis pinnatis incises, racemis simplicibus.* Other species had only one name, but they were whoppers, like *Monolasiocallenomenophyllorum* or *Hypophylocarpodendorum*.

"Barbarian jargon!" Linnaeus called this tangle of nomenclature. With a plant as potentially harmful as poison ivy, it was obviously of some importance to know precisely which plant you were describing. He tossed out all the polysyllabic jumbles and created a simple but elegant plan: everything should have just two names.

The first name is always that of the genus, or family group: *Rhus*, in the case of poison ivy, which comes from an ancient Greek root word meaning "red." *Radicans* is the specific epithet, the part of the name unique to this particular species. Linnaeus was referring to one of poison ivy's most unusual features: the dense, furry-looking rootlets that radiate outward from the stem. (Poison ivy's scientific name has since changed, though; stay tuned.)

Linnaeus was a young man fresh out of college when he began his self-appointed task of classifying the world's plants. Like many recent graduates, he was long on ideas and ambition but short of cash. It was a lucky break for him—and for science—when, during a trip abroad, the penniless young man was introduced to George Clifford, a fabulously wealthy English banker living in Holland. Clifford had a phenomenal garden, with greenhouses filled with specimens from all over the world.

Linnaeus described himself as "bewitched" by Clifford's garden, which encompassed acres of groves, fountains, and flower beds, including state-of-the-art greenhouses, and a private zoo. "The earth's strangest wonders," Linnaeus enthused. "Our beloved Americans, all the progeny of the New World: innumerable cacti, orchids, passion flowers, magnolias and tulip trees." The sight of all these newly discovered plants, including some lovely specimens of poison ivy, made the young naturalist itch to try out his new system. It was a perfect first job for a budding naturalist.

Linnaeus enthusiastically filled thousands of pages with meticulously pressed plants, including several specimens of poison ivy. The specimens were securely fastened to the thick herbarium paper, decorated with an ornate engraving of a Grecian urn.

Botanists since Aristotle had been debating how to classify plants. Some grouped plants by color, or size, or leaf arrangement. Others grouped them by the ways humans used them: medicinal, edible, and so forth. But Linnaeus looked deeper. "Truth ought to be confirmed by observation," he declared, and he closely studied the underlying structure of each plant. He decided that his system of plant classification was to be based solely on the number and arrangement of reproductive organs.

This decision seems unremarkable enough, and of interest to no one but the handful of European botanists who were engaged in a highly erudite and apparently unending debate over taxonomy. But then Linnaeus did something that rocked the scientific establishment. Never one to mince words, the outspoken young naturalist published a little pamphlet that outlined his ideas, describing plants' reproductive parts in words usually reserved to describe human sexuality.

Scratching the Itch:
Historic Poison Ivy

Carolus Linnaeus repeatedly gathered poison ivy, poison oak, and poison sumac specimens, placed them in plant presses, and mounted them on herbarium paper. This, of course, would seem like a recipe for disaster, since the dermatitis reaction is caused by the oily sap of the plant—which emerges abundantly whenever the plant is picked, cut, or damaged in any way. Squeezing the leaves between the lids of a plant press would seem to be the ideal way to express the juice, rather like an incredibly toxic cider press—the perfect way to give yourself a truly historic case of poison ivy.

However, in all his copious writings, Linnaeus never mentions developing a rash. He was no stoic, unhesitatingly recounting his many other ailments, including insect bites, diarrhea, and toothaches, but poison ivy doesn't seem to have been among his woes. Perhaps he was immune.

The urushiol in the plant persists for an incredibly long time. Botanists have gotten a rash from touching specimens more than a century old. Linnaeus's specimens from Clifford's garden, which are still preserved in London by the Linnean Society, are more than two centuries old now, but it's possible they could still contain enough of the toxin to cause a nasty rash.

Flowers, as we all learned in high school biology class, have male and female parts. The male parts, the stamens, hold little clumps of pollen. The pollen must get to the pistil, or female part, of another plant for the flower to develop into a fruit. Linnaeus unblushingly likened pollen to sperm, carried by the stamens or "husbands," who would impregnate the "ovaries" of their virgin "brides."

Linnaeus began with the type of flowers he called Monoclines, which he described as "husbands and wives in the same bed" (in other words, stamens and pistils in the same flower). Then he went on to list the various erotic possibilities with brutal frankness. There were plants called the *Monandria*, with one "husband" decently pairing off with one "wife" (i.e., flowers with one stamen and one pistil). But there was also the class known as the *Diandria*, with two husbands in the same bed with one bride. *Triandria* had three husbands in the "nuptial bed," and so on.

It was really a simple and effective system. Very user friendly. All you had to do was count the stamens on, say, a phlox flower, and instantly you knew that this five-stamened plant was properly placed in the *Pentandria* (five husbands). The system went all the way up to the *Polyandria*, which Linnaeus described as having twenty males or more in the same marriage bed with one female.

It doesn't seem to have occurred to Linnaeus, caught up in the excitement of developing his groundbreaking botanical ideas, that this sort of frank talk was going to ruffle a few feathers. Since Clifford corresponded with virtually all of the leading botanists of the day, Linnaeus's work received a lot of attention. (Some botanists were outraged less by the immorality than by the fact that this young whippersnapper had ignored or discarded their life's work.) Linnaeus was at first unaware of the ripples of shock that began to spread through Europe. Letters and scholarly articles flew back and forth, discussing the news. A German botanist named Johann Siegesbeck was especially outraged. "What man will ever believe that God Almighty should have introduced such confusion, or rather such shameful whoredom, for the propagation of the reign of plants?" he demanded. "Who will instruct young students in such a voluptuous system without scandal?"

Up till this time, botany had been considered a genteel topic appropriate for young ladies. But Linnaeus was using language that would make a sailor blush. He published a pamphlet on his new system, *Systema Naturae*, listing plants like the *Polygamia*, in which the husbands were carrying on not only with multiple wives but with "concubines" (pistils that were sterile and would not produce "offspring" in the form of seeds). Scandalized teachers felt that botany was no longer an appropriate study for young women, whose passions might be overstimulated by this sort of lascivious vocabulary. The ripples of outrage spread beyond the botanical world, as accusations of indecency caught the attention of religious authorities all the way up to the Pope, Clement XIII, who banned Linnaeus's works from the Vatican libraries.

It seems, at first glance, as though Linnaeus was out to shock people on purpose. Why insist on classifying plants based on their sex life? Wouldn't leaf structure do just as well? His X-rated system begins to make sense if you study (as he did) an assortment of poison ivy plants.

Poison ivy, like Cleopatra, is capable of infinite variety. The leaves are compound, which means that they're divided into parts, called leaflets. And they're highly variable. No two poison ivy leaves are exactly alike.

First of all, they can vary wildly in size. This seems to be related to the habitat that the individual plant is found in, whether the area is wet, dry, shady, sunny. For instance, poison ivy plants that grow in densely shaded forests tend to have bigger leaves for maximum photosynthesis. Plants on sunny beach dunes have much smaller leaves.

Then, while there are usually three leaflets on a stem, there can be two. Or four. Or seventeen. The leaves are often smooth but could be fuzzy, hairy, or leathery. It's enough to drive a botanist crazy. So ignoring variables like leaf size and color, Linnaeus concentrated on the flowers. He lumped more than a dozen species in the genus *Rhus*, including many plants that did not have that nasty tendency to cause rashes, basing his classification solely on the structure of the tiny five-petalled flowers. (See appendix for more on poison ivy, oak, and sumac nomenclature.)

As Linnaeus continued his classification of Clifford's garden, he especially enjoyed studying the flowers, perhaps remembering his father's garden that he had loved as a child. Among the jasmine, camellias, and roses that he pasted on his sheets of herbarium paper are the delicate, fragrant blossoms of poison ivy.

Yes, indeed, poison ivy has flowers. They're really quite attractive, a long drooping cluster of dainty, five-petalled blossoms, a pale ivory color. In spite of the fact that they're no less rash-inducing than the leaves, Linnaeus studied them closely and discovered something remarkable. Peering dangerously close to the blossoms—he had such sharp vision that he rarely used a magnifying glass—the young scientist observed that not all poison ivy flowers were the same. He noted that the flowers on

Scratching the Itch:
Oak, Ivy, or Sumac?

Even today people are confused by which plant caused the mysterious rash springing up on their ankles. The names poison ivy, poison oak, and poison sumac are often used interchangeably. But, as Linnaeus made clear, they're separate species, with overlapping geographic ranges. Poison oak is a low-growing shrub found on the east and west coasts of the United States, Canada, and Mexico, with multilobed leaves shaped somewhat like oak leaves. Poison sumac is yet another itch-causing plant, a small tree with white berries and long leaves with many small leaflets.

If you have a rash and you're wondering which of the toxic three you've encountered, don't even worry about it. Doesn't matter, medically speaking. All three contain the same chemical compound: urushiol. The urushiol in the three plants isn't quite identical—there are subtle differences in the chemical bonds—but basically they all affect human skin the same way, so that the symptoms of poison ivy, poison oak, and poison sumac dermatitis reactions are essentially the same. The American Academy of Dermatology lumps the three species of plants together when discussing the treatment of skin conditions resulting from urushiol allergy. (There is much helpful information at www.aad.org.)

Are all three equally nasty? Anecdotal evidence seems to suggest that poison sumac is the most potent, and my own experience agrees with this. I generally get only a mild irritation from poison ivy, but the one time I brushed my elbow against a poison sumac bough was an experience to remember.

one plant were structured very differently than the blossoms on another plant. What could be the reason for this odd difference?

Usually the male and female reproductive structures are found in the same flower—sleeping in the same bed, as Linnaeus put it. But there are a few species of plants, like poison ivy, which have the unusual feature of having male and female parts on completely separate individual plants. In other words, you could say that there are girl and boy poison ivy vines.

Male plants have flowers with five stamens surrounding an infertile pistil—they produce sperm, you see, but no fruit. Females have a blossom with a fertile pistil surrounded by sterile stamens—they can produce fruit if fertilized by sperm. It all sounds remarkably human.

Compared to other plants' goings-on with concubines and three-somes and what-have-you, poison ivy has a relatively chaste marital style. Linnaeus coined a new word to describe this, taken from the Greek: *dioecious*, meaning "two houses." Dioecious plants, including poison ivy, he defined as "Husbands and wives living in different houses."

But *why* does poison ivy have male and female flowers on different plants? To maximize genetic variation, it's particularly advantageous for individual plants to avoid pollinating themselves. To avoid self-pollination, some plants have male and female parts of different heights. In some plants they ripen at different times. But the best way to avoid self-pollination is to have the male flowers and female flowers on completely separate plants. Dioecious plants like poison ivy are considered by many botanists to be the most highly evolved form of flowering plant.

But a male plant might be ten feet away from the nearest female—or ten miles. So how does that work, exactly? How does a male poison ivy vine impregnate the female from miles away? He can't exactly drive over to her house and pick her up. Poison ivy's sex life is completely dependent on the help of pollinators.

The tiny poison ivy flowers are often hidden under the wide-spreading leaves, and are almost never noticed by humans. But the blossoms' pleasant fragrance is very effective at attracting insect pollinators, especially hon-eybees. Dozens of species of bees, butterflies, beetles, and wasps seek out poison ivy's rich nectar and in so doing pollinate the plant.

After two years of pasting flowers and leaves onto herbarium sheets and thinking up plant names, Linnaeus published a book, *Hortus Clif-fordia*, to showcase the results of his efforts. The artwork in the frontis-piece, designed to his specifications, shows what Linnaeus thought of his own work—it shows the young naturalist robed as Apollo, giving off rays of light and bringing the gleam of knowledge to the ignorant

Scratching the Itch: Have a Sip of Poison Ivy?

If you've ever stirred honey into your tea, you've likely used honey made with poison ivy nectar. Fortunately, the toxin that is responsible for the rash is not present in either the pollen or the nectar of the poison ivy flowers, so the honey poses no health threats.

world. Linnaeus was absolutely convinced his system was the last word in botanical nomenclature.

But sadly, he was overoptimistic. The wrangling over the correct classification of poison ivy continued, and no one has had the last word to this day. Not long after the publication of *Hortus*, Linnaeus himself decided to rename the plant, changing the name from *Rhus radicans* to *Rhus toxicodendron*, reasoning (perhaps from experience?) that the most important quality of the plant was not its radiating root system but its toxicity.

In 1757, John Ellis, an English proponent of Linnaeus's "sexual system," and Phillip Miller, a famed but old-fashioned London gardener, had a bitter debate over the classification of poison ivy that involved dozens of European botanists. The argument was never really settled. In the twentieth century, scientists began to argue that poison ivy and its irritating relatives should be in a completely separate genus from the harmless *Rhus* species like the staghorn or smooth sumacs, and by the 1970s, poison ivy and its more toxic relations were booted out of the *Rhus* genus and given their own new genus, aptly christened *Toxicodendron*, from the Latin for "poison plant." So poison ivy is now officially referred to as *Toxicodendron radicans*, although older reference books and also some British or Canadian sources still use *Rhus toxicodendron*.

All this botanical hairsplitting can drive you a little crazy. And what does it all matter, really? A *Toxicodendron* by any other name would still itch, after all. Does it really matter how—or if—we keep on lumping and splitting species, endlessly arguing what to call plants?

Linnaeus lived in a time when the natural world seemed inexhaustible, with no end to the dazzling wealth of new plant and animal species waiting to be discovered. But he knew, even then, that names matter. "If you do not know the names of things," he wrote, "the knowledge of them is lost, too."

In the twenty-first century, research into biodiversity has become a hot topic. The mission that Linnaeus embarked upon in the eighteenth century has new urgency today. In spite of the work of Linnaeus and countless other scientists, there are an unknown number of species still waiting to be discovered. Some may go extinct before humanity even notices them.

"We know less about life on earth than we know about the surface of the moon," the ecologist Edward O. Wilson wrote in 1999. "Taxonomy is at the base of biodiversity studies for the simple reason that if species cannot be identified they cannot be studied or marked for preservation." The more we know about the incredible diversity of species that grace our planet, the better able we are to protect that diversity.

More than two centuries ago, Linnaeus understood that. He spent the rest of his life promoting his system of classification. Hypersensitive to criticism, he was deeply offended by all the controversy and maintained a long and bitter feud with many of his colleagues. Siegesbeck in particular never got over the "loathesome harlotry" that plants were up to, and continued to protest. But Linnaeus never apologized. A medical doctor, he saw no reason not to use descriptive, forthright, easily understood words like *sperm* and *ovary*. He eventually prevailed, and his ideas, though much modified, are in worldwide use more than two hundred years after his death.

Linnaeus often named plants after colleagues, choosing the name based on some quality in their personality. He named *Tillandsia*, a genus of plants that apparently survives without water, after a botanist named Elias Tillander, who detested the ocean after a near drowning. A student of Linnaeus's, Peter Forsskål was stubborn and determined, and after his tragic early death Linnaeus named an especially hardy plant *Forsskaolea tenacissima.* But in spite of his poisonous dislike of Siegesbeck, he didn't name poison ivy after his hated rival, choosing instead to label an inconspicuous weed *Siegesbeckia.* To Linnaeus, poison ivy was a powerful, handsome plant—and clearly one with a big future.

CHAPTER 5

The Biggest Book

It's quite possibly the biggest book in the world. John James Audubon's classic *Birds of America* is a whopper, each page three feet tall and two and a half feet wide. A complete four-volume set of the first edition, known as the "double elephant," weighs two hundred pounds. It's hard to read in bed.

Publishers insisted that such a monster would never sell and urged Audubon to scale his ambitions down a tad. But Audubon wanted his book to represent the magnificent nature of North America—life-size—and he wouldn't settle for anything less.

And as it turned out, he was right. The enormous book sold for equally enormous sums, even in 1837, when it originally appeared. In 2000 a first edition was auctioned off to the sheikh of Qatar for over eight million dollars. A big book in every way.

Open the book's ponderous cover, and inside you'll find the American wilderness in all its majesty: shapes, colors, dramatic and graceful forms, painted by a master. Yes, on these immense pages, John James Audubon has depicted hundreds of the most beautiful and important species of native American plants.

That's right, plants. The book is filled with gorgeous renderings of plants, from tiny mosses to *Magnolia grandifolia*. Turn to Plate #132 to admire a particularly lovely portrait of poison ivy.

Oh, yes, there are a few birds in the book, too.

No one ever notices plants. They're always the backdrop to the action, the stage on which events take place. Botanists (who are more than a little miffed about this) have given a name to this common human tendency to ignore plants: green blindness. The term was coined in 1998 by two botanists, Elisabeth E. Schussler and James H. Wandersee, who also called it "plant blindness" to distinguish it from literal color blindness. They defined it as "the inability to see or notice the plants in one's own environment, leading to the inability to recognize the importance of plants in the biosphere and in human affairs."

But it's not an inability so much as a habit—a very ancient habit. It's part of our genetic hardwiring. Humans are designed to respond to things that look like us—things that have two eyes, a nose, and a mouth. There's a strong human tendency to see a face even in inanimate objects (glance at an electric socket, for example). Plants just don't register as animate.

The human brain can't process every single one of all the millions of things it can see in a glance. So the brain, unconsciously, picks and chooses what to focus on. When we scan a location, our inner caveman looks for threats—sudden movement, a gleam of color, anything out of the ordinary. As Wandersee says, "Plant blindness is the human default condition." The plants just blend into a green blur, a sort of botanical wallpaper.

This, by the way, is the secret of why it's so easy to wander into a patch of poison ivy. I can't tell you how often I've strolled into a patch of thistles, or nettles, or PI, my mind on higher things, blind to the danger until too late.

John James Audubon was no different from the rest of us in focusing primarily on movement and color. What's more animate, more fascinating to watch, than a bird? He grew up on the island of Saint-Domingue, now called Haiti, and gazed entranced as rainbows of birds fluttered through a background of tropic greenery. His father, fearful of a slave revolt on the turbulent island, eventually shipped the boy off to the less scenic but safer habitat of the family home in France, and then eventually to Pennsylvania. But the young Audubon continued to be fascinated—obsessed—with birds. "As I grew up," he remembered, "I was fervently desirous of becoming acquainted with nature . . . instead of applying closely to my studies I . . . usually made for the field, my little basket filled . . . to bursting with nests, eggs, lichens, flowers, and even pebbles from the shore of some rivulet by the time I came home at evening."

As he grew up, he had trouble, as so many artists do, settling into a humdrum profession. He dabbled in business ventures, failing at them all. His greatest joy came from ambling through the woods and fields, drawing what he saw there. Finally, he gave up on mundane pursuits. Like John Smith, Tradescant, Bartram, and many others before and since, he found that nature, dangers and all, was an addiction he couldn't resist. He set off on an odyssey through the wilds, his goal an ambitious one: to paint every single species of North American bird.

Quite possibly he explored more of wild America than any other voyager of his time. He travelled from the Dry Tortugas to Labrador, from the Atlantic coast to the Mississippi, determined to create pictures that would show the world how magnificent the bird life of America truly was. But to paint birds, first he had to find them.

Contemplate the immensity of his task, fellow birders. There were no high-powered binoculars, no spotting scopes in those days. There were as yet no reference books, no field guides in existence to help him in his quest. No one, in fact, even knew how many species of birds there were. But in spite of ditching all that school in his youth, Audubon had learned to read nature like a book. "I know I am not a scholar," he wrote, "but meantime I am aware that no man living knows better than I do the habits of our birds."

During the thousands of hours he spent in the field, searching out the species he wished to portray, he formulated what he called the Ornithologist's Rule: "The nature of the place—whether high or low, moist or dry, whether sloping north or south, or bearing tall trees or low shrubs—generally gives hint as to its inhabitants."

So he knew that cattail marshes were the place to go for red-winged blackbirds and great blue herons. For crossbills he needed hemlock groves; he sought out grasslands to sketch meadowlarks, vine-covered shrubby edges for catbirds. And when he wanted to draw woodpeckers, he went to poison ivy.

It's no accident that his painting of the three-toed woodpecker shows the birds cavorting on leafy poison ivy vines. Audubon must have been aware, from his years of observation, that birds in the woodpecker family—downy, hairy, pileated, yellow-bellied sapsucker, common flicker, etc.—are especially fond of snacking on the small, whitish fruits of poison ivy.

Given the technology of the time—no videos, no cameras—there was no practical way to paint birds from life. Disturbing as it seems to us today, Audubon was an enthusiastic hunter who personally shot each bird, then arranged it in a lifelike pose and mounted it on wires. (After sketching his specimens, he occasionally would dine on them.) Audubon could have painted his birds—as did all other scientific illustrators of the time—posed stiffly against a blank background, but he chose to paint in a stunningly new way. As though he had realized that photography would someday exist, he painted his birds like a candid snapshot. His goal was "to represent nature . . . to copy her in her own way, alive and moving!" He depicted birds on the wing, in the act of hunting, eating, or nesting, and he showed them among the plants where he found them.

From buffalo grass to tulip trees, he re-created the flora of North America in all its stunning beauty. In fact, as you turn the giant pages, you'll find picture after picture in which the birds are all but overshadowed by the plants. In the engraving of poison ivy, the vine is as colorful and vigorous as the woodpeckers are; plant and birds are clambering up the branch of a dead tree, side by side. In the engraving of the species then known as Bonaparte's flycatcher, it's hard to even spot the flycatcher at first—the tiny, gray bird is dwarfed by the magnificent scarlet seedpods of a magnolia tree. Ruby-throated hummingbirds are all but hidden among huge trumpet-flower blossoms.

Audubon understood plants' key role as shelter, and emphasized it: the intricate tangle of grasses that make up the marsh wrens' nest is more astonishing than the drab birds themselves. When Audubon painted his enormous image of a female wild turkey, he included a life-size thicket that securely camouflaged her young. Drawing after drawing shows the birds interacting with plants—birds eating seeds, birds sipping flower nectar, birds perching, nesting, hiding in, and feeding on plants.

Audubon knew, right from the start as he painted his vultures and buntings, ducks and warblers, that he was creating a new kind of book. Without perhaps being aware of it, he was also beginning the creation of a new kind of science, by portraying in his artwork the subtle and tangled web of relationships that exist among living things and their environment. The study of these interrelationships grew to a full-fledged discipline in the twentieth century, the science known as ecology.

As Audubon perceived, poison ivy is a plant of enormous ecological value. Its berries are eaten not only by woodpeckers but by dozens of other bird species as well. But there are countless other factors that make poison ivy valuable to wildlife—all those complex threads in the web that we don't see at first glance.

First of all, poison ivy grows in an enormous range of habitats: from the chilly spruce forests of Canada to the desert scrub along the Rio Grande, on Appalachian mountaintops, in the piney woods of Georgia or along the beaches of Cape Cod. No matter where Audubon travelled, he could be sure of finding poison ivy somewhere nearby.

Poison ivy is a plant that comes in many shapes and sizes, with a wide range of growth patterns, depending on its habitat. On the forest floor, poison ivy's habit of growing as an ankle-high ground cover creates shelter for small animals like slow-moving box turtles. The broad leaves act as umbrellas, creating crucial shade for amphibians like toads and salamanders, who must maintain moist skin in order to breathe.

Or poison ivy can be a high-climbing vine bearing fruit all along its length, and so enabling a vast diversity of wildlife to take advantage of it. There might be tender leaves six inches above the ground, in easy reach of passing rabbits. In winter, starving deer could nibble nutrition-packed twigs and buds growing four feet high, poking out above the snow. Or poison ivy might have clusters of berries a hundred feet in the air, available to every passing chickadee and mockingbird.

As Audubon observed, poison ivy plays a role in the nesting of many bird species. I've seen it myself—right in my backyard is a big old poison ivy vine, with sturdy stems branching out at right angles and providing a broad, stable platform for a blue jay's nest. Cardinals actually use soft, fuzzy poison ivy rootlets to line their nests.

And then there are the bugs, the basis of the food chain for countless animals. Hundreds of species of insects use poison ivy for food and shelter. Larval insects spin silk to roll and fold the leaves of poison ivy to enclose themselves, making a cozy spot for cocoons. Poison ivy leaves are a popular insect snack. Audubon's painting of poison ivy and woodpeckers, in fact, illustrates this vividly. It's not just a picture of birds on a decorative vine. If you look closely you'll notice that almost every poison ivy leaflet is well chewed, marked and scarred by the gnawing of insects. And all these insects, of course, are food for the woodpeckers as well as many other avian species, and other predators like spiders, toads, and salamanders.

The "ornithologist's rule" that Audubon had figured out was true, and then some. The "nature of the place" doesn't just give a "hint" as to its inhabitants—the nature of the place dictates its inhabitants.

John James Audubon finally emerged from the wilderness with a monster portfolio containing his huge paintings. With high hopes, he went to Pennsylvania and unveiled his work at the Philadelphia Academy of Natural Sciences. But his enormous drawings, created with such diligence and love, were met with scorn. The scientists scoffed at his lack of academic credentials. He was criticized for not arranging the birds in order of the now-popular Linnaean system. While a few members of the prestigious society did appreciate his skill at drawing, most were baffled by Audubon's new, dramatic style of art. And even those who admired his elegant birds were put off by his realistic depictions of plants, complete with insect holes and spiderwebs. He'd even included rattlesnakes, dead rats, and all sorts of unsightly things. George Ord, a wealthy and influential founding member of the academy, was completely unimpressed. He called on his fellow academicians to reject this "impudent pretender and his stupid book."

Audubon's worst sin, especially derided by Ord and the other scientists who studied his paintings, was that he had mixed zoology with botany. This was deemed ridiculous, confusing, and in fact, counterproductive. Plainly, they should remain two completely distinct studies.

The learned members of the prestigious academy voted overwhelmingly against admitting Audubon as a member.

It wasn't till Audubon crossed the Atlantic that he hit pay dirt. A consummate showman, he realized that he had to sell himself as well as his drawings. "Erect, and with muscles of steel," the backwoods artist was "quite a handsome figure . . . with a fine set of teeth." (The less-than-modest description is his own words.) He wore his buckskins, with his "luxuriant" ringlets flowing under his coonskin cap, and regaled rapt Londoners with tales of narrow escapes from Indians and rattlesnakes. His experience in the Old World was not unlike that of Pocahontas, or of poison ivy itself—Europeans found themselves enthralled by the strangeness, the wildness, and the hint of danger of these strange imports from the Americas.

In Great Britain he quickly found a publisher willing to take on the monumental task of turning his unwieldy portfolio of artwork into a book. Each incredibly detailed painting had to be engraved on copper plates. Then each engraving was painstakingly hand-tinted by a team of artists, with an anxious Audubon overseeing every step of the process. The massive "double-elephant" pages were then bound together in a series of weighty volumes. Although the publisher begged him to make the volumes smaller, cheaper, and more manageable, Audubon wouldn't budge. "How could I make a little book," he demanded, "when I have seen enough to make a dozen large books?"

John James Audubon is credited with depicting more than four hundred species in his famous masterpiece. But of course, that's species of *birds*. His big book includes much more than that—hundreds, perhaps thousands, of identifiable species of plants, each leaf vein and flower petal delineated with the same precision as the birds' plumage. Credit, by the way, should also go to Joseph Mason, a young art student who accompanied Audubon on many of his travels. Mason did a lot of the grunt work, carrying art supplies and such, but he was a talented artist in his own right. He is credited with having painted the plant life in at least fifty of the paintings chosen for *Birds of America*.

Audubon's avowed goal was to paint birds. But having spent countless hours observing them, he couldn't help but notice the plants that sustained them. He overcame "green blindness" to show the world the diversity of plants that were the source of America's incredible wildlife. John James Audubon had created a much bigger book than he knew.

CHAPTER 6

Strong Medicine

Valenciennes, France, 1780

JUST AN ORDINARY GREEN PLANT. THREE LEAFLETS. NOTHING SPECIAL about it. Surely the professor's warning couldn't be serious?

Professor Andre Joseph Dufresnoy, a well-known French physician, often gave lectures to his medical students using specimens of plants from his "physick garden." But one of his students couldn't believe all the ridiculous claims his teacher was making about the little plant called *Rhus radicans*. (At this time, the earlier nomenclature was still common.) True, *Rhus* was an exotic plant from America. But could ordinary-looking leaves like this really attack people, scorching and blistering the skin? *Ha.*

The youth picked up the specimen of *Rhus* in his bare hands. The professor advised him to put it down. But as all teenagers know, advice is for cowards. The young man smilingly rubbed the plant all over his face, hands, and arms. The rest, as they say, is history.

Dr. Dufresnoy would have been more than human if he didn't murmur "I told you so" when, two days later, the know-it-all student came down with a terrible rash. At any rate, the good doctor recorded every symptom of the dermatitis reaction in excruciating detail. The overconfident student must have been acutely allergic, because he developed an epic rash that covered his face, swelled his eyes shut, and "spread all over the body, chiefly the hairy scalp and the privates, which he tore to pieces with scratching."

But after a few weeks, the humbled young man visited his teacher with an amazing tale. The rash had finally vanished—and so had a painful and chronic case of skin sores (probably herpes) from which he had suffered for more than six years. The patient claimed that after he had tried every treatment medical science of the time could offer, his skin condition had been completely cured by a brisk application of *Rhus*, also known as poison ivy.

Dr. Dufresnoy was intrigued. Could *Rhus* be the miracle cure-all every doctor longs to find? He gathered more of the plant from his garden and set to work.

Dufresnoy, a highly respected medical man and chief physician to the military hospital at Valenciennes, was not alone in his interest in plants. Every medical student of the time was well acquainted with botany, and many doctors grew their own physic gardens full of medicinal herbs. Housewives, of course, also cultivated herbs, using them for gentle household remedies that had been known for millennia: catnip tea for insomnia, mint for upset tummies, dandelion as a spring tonic. But medical practitioners like Dufresnoy were seeking plants that had muscle.

In the eighteenth and nineteenth centuries, the prevailing medical opinion was that to do any good—to really conquer a serious illness—you had to get out the big guns. Drastic, dramatic remedies were the solution—kill or cure. And indeed, remedies were often so severe that the patient died of them: the removal of quarts of blood through an incision in a vein, for example. Hefty doses of poisons like arsenic or belladonna. Pills made of mercury mixed with a tasty dose of lead. Electric shock therapy. Rubbing a little poison ivy on yourself seems almost tame by comparison. Treatments that had a devastating effect on the patient were known as "heroic" cures, and Dufresnoy wanted to learn if *Rhus radicans* had the potential to be a hero.

As many conscientious researchers do, he began his research by experimenting on himself. He drank an infusion made from a single leaf of poison ivy and waited to see what would happen. When there was no ill effect, he tried taking two leaves. Fortunately for him, he was not acutely allergic to urushiol, the irritating chemical found throughout the plant, and was able to gradually up the dosage. He discovered that he could ingest up to twelve leaves before they caused him mild internal distress.

He then experimented on volunteer patients with various skin conditions. With gathering excitement and confidence he reported tremendous success in relieving skin sores, pain, and ulcers.

Dufresnoy ignored the skeptics who warned him that *Rhus* was too dangerous a plant to meddle with, and advocated poison ivy as a highly

Scratching the Itch:
The Fortunate 15 Percent

When reading of how someone like Linnaeus, Tradescant, or Dufresnoy handled poison ivy with impunity for so long, it's impossible to keep from wondering: how did they do it? They may well have been among the approximately 15 percent of humans who are immune to poison ivy. No one knows why some people are immune, though it does seem to be somewhat hereditary.

However, appearances can be deceiving. A person can literally roll in poison ivy time after time, convinced of their invulnerability, and then after years of repeated exposure finally develop the full-blown allergic reaction. But some people do maintain a lifelong immunity, and that may explain the enthusiasm of some medical practitioners, and some patients, for using poison ivy as a medication.

Dufresnoy almost certainly was the fortunate possessor of an immune system that simply did not react to poison ivy. He gathered the leaves, pressed them, distilled them, drank them, rubbed them on his skin, time and again. Before giving a tincture of concentrated *Rhus* leaves to the boy who was his first paralytic patient, the brave doctor wrote: "I took some of it myself, to ascertain if it was poisonous; I found, however, that it was perfectly innocuous, in considerable doses." Anyone who could pen words like these about poison ivy must have had a high degree of immunity.

effective medication. He continued to have such good results curing skin diseases that his reputation spread. One day a young man suffering from hemiplegia, or paralysis of one side of the body, came to Dufresnoy and implored his aid in curing the paralysis, which had occurred after an attack of epilepsy. Dufresnoy was willing to try his favorite remedy, and it apparently worked. The cure was so spectacular that many more sufferers came to him. He published books and articles in scholarly journals announcing that poison ivy was a sovereign remedy for many forms of paralysis.

Other doctors began to use *Rhus* to treat paralysis symptoms. The most popular method was a lotion made from poison ivy well rubbed into the skin, which seemed to be especially effective in curing paralysis in the extremities caused by rheumatic fever. Admitting that poison ivy's effect on the skin could be unpleasant, doctors recommended opium, arsenate of lead, or mercury to take the sting away from the side effects while *Rhus* was doing the good work of curing the ailment.

It's tempting to speculate that, if the paralysis was psychosomatic, a good dose of poison ivy rubbed on the skin might inspire the patient to

Healing Mysteries

Thousands of years before Dufresnoy grew medicinal plants in his physic garden, the ancient Celtic peoples who inhabited his native France were well versed in herbal medicine. A Celtic legend tells of Dian Cecht, god of healing, who made a silver hand for a warrior who had lost his hand in battle. But Dian Cecht's son was an even more powerful healer, and he created a new hand made of flesh. Upstaged, the angry god killed his own son in a jealous rage.

Green plants sprouted on the son's grave, and his loving sister gathered them and spread them on her cloak in order of their medicinal strength, so that she could share them with mortals. But the furious Dian Cecht flung aside the cloak, forever scattering the herbs so that no one would know their secrets. To this day, the healing plants keep their mysteries.

move around a bit. But case after case was listed in textbooks, not only by Dufresnoy, but by other doctors who reported excellent results in a variety of cases. And so began the controversy that has lasted to this day: is poison ivy good medicine?

There's a kind of sense to the idea that a plant that has an enormous effect on the body is powerful enough to affect disease as well. The lovely wildflower foxglove, for instance, yields a drug called digitalis, which taken in large doses can cause a fatal heart attack. Taken in small doses, it can be a lifesaver. Its medicinal benefits were first described in 1785, and an extract of the plant is still used today to control irregular heart action.

Although we call it *poison* ivy, the irritating element in the sap is itself not a poison. Urushiol isn't toxic in the same way as, say, arsenic, which would kill any human who ingested enough of it. Urushiol is an allergen, like cat dander, or peanuts. Some people can cuddle up to a fuzzy Persian; others are sent running for their inhalers when a cat walks in the room. Some people love peanut butter sandwiches; others can die if they ingest even a tiny speck of a peanut. Every individual reacts to allergens differently.

When a harmful disease invades your body, your immune system battles it. In your body are millions of cells called effector T-cells. They patrol the body, looking for invading cells and viruses, like vigilant night watchmen. But sometimes your immune system is too vigilant. A harmless visitor (urushiol) is seen as a deadly threat. Your own body unleashes weapons to stop it. The immune system kicks into high gear, resulting in the symptoms of urushiol allergy: "rednesse, itchynge, and lastly blisters," as Captain Smith put it.

It was Dufresnoy's theory that this violent kick start could be a good thing, revving up the patient's immune system to a point where it also battled undesirable symptoms like herpes sores. And Dufresnoy wasn't a lone crank. Many people agreed with him.

Dr. Jacob Bigelow's 1817 *American Medical Botany* was a classic reference book of its time. Dr. Bigelow, a Harvard medical professor, described poison ivy's successes in medicine, noting with a touch of awe that the plant could "afford a more violent external stimulus than any medicinal substance with which we are acquainted." But he concluded that the plant was "too uncertain and hazardous to be employed in medicine." Dr. Charles Millspaugh, a prominent American physician, hedged his bets by describing both poison ivy's medical benefits and cures for the rash in the same chapter in his textbook.

Dr. John Scudder, author of the respected 1870 medical text *Specific Medication and Specific Medicines*, devoted many pages to praise of *Rhus*. Dr. Scudder wrote that "The *Rhus* has proven one of our most valuable medicines and will be highly prized by every practitioner once its use is learned . . . I have seen the remedy relieve the severest pain in an hour." *Materia Medica and Therapeutics of the Vegetable Kingdom*, written by Dr. Charles Phillips in 1879, had high praise for the benefits of *Rhus* (for some odd reason, physicians never referred to it as poison ivy) and quoted Dufresnoy extensively.

Like many remedies of the day, *Rhus* was prescribed for a host of symptoms, including inflammations, ulcers, paralysis, headaches, and syphilitic infections. Poison ivy was listed for decades in the *Official Pharmacopeia* of Great Britain and the United States, and was still included in *King's American Dispensatory* in 1898, more than a century after Dufresnoy's cocky student made his historic mistake.

Interestingly, Native American traditional medicine had for many centuries been on the same track. A classic work titled *Iroquois Medical Botany*, a compilation of oral tradition from Iroquois healers, frequently suggests poison ivy as a remedy for skin rashes. Gladys Tantaquidgeon, an anthropologist and traditional Mohegan tribal medicine woman, noted that poison ivy has "valuable medicinal properties," adding that "a poultice is made from the roasted crushed root." The ethnobotanical records of many Native American tribes refer to poison ivy as a medical plant, especially effective in treating skin conditions.

Poison ivy to this day is a popular remedy in homeopathic medicine. The idea behind homeopathy is that "like cures like," so that a substance that causes symptoms in a healthy person will cure similar symptoms in a sick person. The concept became popular in the late 1700s and is still widely practiced, especially in Europe. Homeopathic practitioners use tiny amounts of urushiol, much diluted. A tincture of the fresh leaves is mixed with water and is used to treat skin conditions, and also as a pain reliever. On one homeopathy website, *Rhus toxicodendron* (many homeo-pathic practitioners use the older scientific name) is listed among the top cures most useful for a wide range of ailments. Deborah Olenev, a California homeopathy practitioner, writes, "If I were told that I could limit my home remedy kit to only five remedies, I would be sure that *Rhus toxicodendron* . . . were one of the five." Like Dufresnoy, she lists case histories, and adds, "I prescribe this remedy to more clients than any other."

—◆—

Andre Dufresnoy was not the only, but perhaps the greatest, admirer that poison ivy has ever had. He spent much of his life tirelessly researching and experimenting on the fuzzy vines and three-parted leaves. He wrote the first—and only—book ever published on the medical benefits of *Rhus*, staking his professional reputation on the efficacy of his favorite plant.

But his poison ivy research came to an abrupt halt in 1793, when the French Revolution brought turmoil to his country and changed his life forever.

As the luxury-loving French monarchy toppled, chaos and confusion spread across France. A Reign of Terror began. Anyone suspected of being opposed to the revolution was in danger of immediate execution. France's nervous neighbors, Prussia, England, and Russia, soon began to threaten invasion, and the Revolutionary government mobilized huge armies. Dufresnoy was summoned from his quiet physic garden to become the physician-in-chief of the newly formed Army of the North.

But the conscientious doctor shortly fell afoul of the hypersuspicious authorities, who feared disloyal folk scheming to reinstate the monarchy. He made the dangerous mistake of treating an enemy patient (we don't know if he prescribed poison ivy) and was immediately fired from his post, accused of the crime of "an act of humanity to a royalist." Dufresnoy was now a marked man.

The doctor happily returned to his peaceful garden. He resumed his enthusiastic cultivation of poison ivy and frequently sent seedlings of the plant to friends and colleagues. After dispatching one such sample, he wrote to the recipient inquiring tenderly after the health of his beloved specimens. "How are our dear Rhus?" he wrote in his letter. "How I long to see them!"

Unfortunately, he didn't make it clear in his letter that he was referring to a plant. *Russe* in French means "Russian." Already under suspicion, his correspondence monitored, this was pounced on as clear evidence of treason. He was denounced to the Revolutionary Tribunal for instigating a "traitorous correspondence with the Russians"—an offense punishable by death.

Dufresnoy was arrested and brought up before the bloodthirsty tribunal, which had already condemned tens of thousands to death under the razor-sharp blade of the guillotine. In spite of his efforts to encourage the use of poison ivy, *Rhus* was still far from being a household word, and the unfortunate doctor must have despaired that any nonbotanist would believe his unlikely story. Providentially for him, the French Revolution was shaken by an internal rebellion; the bloodthirsty leaders of the Reign of Terror were swept from power, and Dufresnoy was released from his dungeon in the nick of time.

Dr. Dufresnoy joyfully returned to his medical practice in Valenciennes, but died only a few years after his narrow escape. Immediately after his death his family dug up and destroyed every *Rhus* specimen in the garden. Their reasons are unrecorded, but probably his relatives did not share his immunity and were simply tired of dodging itch-inducing foliage every time they walked down the garden path. The extermination of his most prized medication would have been seen as a tragic act of vandalism by poison ivy's greatest fan, but the eradication of *Rhus radicans* from the physic garden was perhaps symbolic of poison ivy's bleak future prospects in the field of medicine.

But poison ivy had other worlds to conquer, other niches to fill. The height of its fame was still to come.

CHAPTER 7

Royal Color

Paris, France, 1800

JOSEPHINE DE BEAUHARNAIS WAS THE UNDISPUTED LEADER OF THE swankiest high society in eighteenth-century Paris. Married to the most powerful man in the world, Emperor Napoleon Bonaparte, the elegant empress spent money like water, on clothes, on jewels—and especially on plants.

She squandered huge sums beautifying her beloved estate of Malmaison, a verdant retreat just outside of Paris: building greenhouses, designing archways and fountains, and creating magnificent ornamental gardens. Poring over the catalogs of garden suppliers and nurseries, she spared no expense, introducing to France new and unknown plants like camellias, phlox, and jasmine. She imported rare tulips and expensive lilies and was especially obsessed with roses. She even splurged on poison ivy.

Poison ivy's brilliant gold-and-purple fall coloring has always gone especially well with royalty. Indeed, the now-despised plant has a regal history, starting long before Josephine invited the colorful vine into the rose garden at Malmaison. Poison ivy was not an unusual sight at the most magnificent palace of them all.

Versailles. The very word embodies royalty, ostentation, wealth, luxury. It's interesting, I think, that the name actually came from a word that means "to turn the soil." In fact, the first Versailles was a humble farming village on a low hilltop, surrounded by a patchwork of crop fields that the French peasants had painfully scratched out of the gravelly soil. But King Louis XIII, a passionate fan of stag hunting, began using the forest that surrounded the fields, conveniently not far from Paris, as a hunting spot. Even King Louis knew enough about ecology to realize that you can't kill more than a thousand stags a year without some habitat to support them, so he added more and more acreage to the royal estate to ensure an adequate supply of deer to hunt. The original peasant farmers were politely asked to leave, and a mansion was built at Versailles to serve as a royal hunting lodge.

Louis XIII's heir, Louis XIV, wasn't such an avid huntsman as his father had been. He wanted to turn the hunting lodge into a vast and imposing palace that would became the seat of the French government—a worthy home for the ultimate monarchy. He began to plow serious money into Versailles, especially the gardens and pleasure grounds around it. Again, the peasants weren't consulted—whole villages were relocated if the farmers' huts marred the view from Louis's windows.

In keeping with the fashion of the time, the royal gardeners weren't aiming for a natural look. They wanted to subdue nature, to impose human order on wildness. Hills were flattened into terraces, running brooks turned into ornamental fountains. Meadows were transformed into lawns manicured to the inch. The gardens were laid out with geometric precision, with straight-angled walls and pathways, and hedges forming complicated knot patterns. Shrubs were shorn and twisted into fantastic shapes and figures. Trees were planted in straight lines and trimmed at a uniform height. Each plant was set apart from its neighbors, like caged animals in a zoo. Exotics from all over the world graced the garden, and poison ivy was one of the many foreign plants King Louis admired in his new domain.

The grounds of Versailles became the epitome of the classic French style. The controlled and regimented gardens symbolized the monarch's total control over his subjects and were a horticultural testament to the mastery of man over nature.

But poison ivy is a plant that's hard to master.

Clipping the branches, trimming its untidy foliage, and coaxing the fuzzy vine to run in prearranged patterns would be a hazardous task indeed. Many royal gardeners must have come off on the losing side in a battle with poison ivy. Yet the irritating plant is repeatedly listed in the royal garden inventories of Versailles, appearing as early as 1759.

But why? Colorful autumn foliage is a pleasing sight, few would disagree. But what on earth was so special about poison ivy that would justify working with a plant that could lead to such extremely unpleasant consequences?

For us in the twenty-first century, it's hard to understand the intense enthusiasm with which vivid plants like poison ivy were hailed in the early eighteenth century. The garden in those days was a green place, but not the bright shout-out of color it is today. Exotic species from the Americas, Asia, and Africa were just beginning to make inroads into European gardens. Today, anyone can buy a flat of hot-pink petunias for $2.99, but back then there were no cheap, easy blasts of color like petunias—no zinnias, marigolds, or geraniums available for any money. Peonies, poppies, pansies—none was yet for sale.

Fall and winter in Great Britain and northern Europe tended to be dull affairs. Few native trees and shrubs showed much in the way of autumn color. Summer's green leaves wilted, turned brown, and fell off, ushering in the gray clouds, snow, and mud of winter.

So when European colonists experienced their first North American autumn, they were thrilled by the leafy fireworks: orange and gold maples, scarlet-tinted oaks and Virginia creeper, purple ash trees, and a rainbow of poison ivy. All this was a heady sight to gardeners accustomed to grayish-green Octobers. Poison ivy's colors were deemed worth risking a rash for.

Poison ivy must have been a magnificent sight when clipped and trimmed and displayed in an artistic fashion. Few people today are willing to take on the challenge, but several bonsai masters have made beautiful and striking bonsai using poison ivy. A nursery in Kyoto, Japan, advertises poison ivy bonsai for sale at stiff prices. Nick Lenz, a well-known American practitioner of the ancient art of bonsai, notes that "the species has great charm and interest, especially when fruiting."

But undoubtedly the most famous inhabitant of Versailles was not poison ivy, but the doomed and beautiful queen, Marie Antoinette. Arriving at Versailles as a teenager, the youthful bride of the heir to the throne, she was immediately bored and frustrated by the regimented life of the court, which was rigidly controlled by persnickety rules of etiquette. Soon the impulsive princess was itching to escape from the manicured gardens and live a more natural life. Her indulgent husband, Louis XVI, gave her an estate known as the Petit Trianon, within the grounds of Versailles, for her own personal use.

Marie Antoinette adored plants and ordered rarities from all over the world. Dozens of species of American plants graced her bower at Versailles. (The inventory lists for her garden did not survive the Reign of Terror, but it's highly probable that poison ivy was on them, as the plant was still common at Versailles shortly after the Revolution.)

Most of the brave souls who have gardened with poison ivy have felt that it's better to leave it alone and let it run wild as much as possible. Left to itself, poison ivy will clamber up trees, as well as spread along the ground. This pattern of growth fit in perfectly with Marie Antoinette's preferred style of gardening—the uncorseted, natural look.

Always a leader of fashion, the carefree queen loved the newest type of garden that was becoming all the rage, known as the "Anglo-Chinese" style. Inspired by naturalistic Oriental gardens, English gardeners had embraced an apparently carefree style of gardening—which was actually no less planned and regimented than the French geometric style. Echoing the growing push for freedom that was sweeping Europe, freedom was invading gardens as well.

Gardeners began to take down walls. Trees and hedges were no longer trimmed into cubes and spheres, but were allowed to grow in natural forms. Flowers intermingled, and shrubs were loosely arranged in fragrant groves where vines draped themselves casually over the branches. But nature wasn't really left to run free and wild—it was artfully arranged to look as if it was running free and wild. European gardens mimicked the wilderness of North America.

Scratching the Itch:
Sweet Remedies

Marie Antoinette loved flowers, and many of the fragrant and colorful blossoms in her gardens were potential remedies for a case of poison ivy. There is no record of the queen herself getting poison ivy, but if any of her unfortunate gardeners came down with a rash, the essential oils of several of her favorite flowers could have served as gentle skin soothers. Oil of geranium was, and still is, used for many types of skin irritations. Lavender is another age-old cure for itchy skin and has mild antibacterial properties as well as its soothing scent.

Bear in mind, though, that oils should not be used on the skin immediately after exposure to poison ivy, as they may spread the urushiol around. Oils also should not be used to cover a weeping rash. Rather, the essential oils of flowers are a pleasant way to soothe slightly irritated, reddened skin and help the healing process. (See appendix for more ways to soothe the itch.)

Marie Antoinette and Louis XVI did not survive the French Revolution, but poison ivy did. As a symbol of the hated monarchy, Versailles was nearly destroyed by furious mobs at the height of the Terror. But as time passed and things calmed down a bit, the palace grounds were opened to the public and turned into a park. Commoners flooded into the once-forbidden royal enclave to admire King Louis's knot gardens and topiary and Marie Antoinette's bowers, grottoes, and wildflower meadows. Housewives scrubbed the family laundry in the gilded fountains and spread their clothes to dry over ornamental shrubs.

As the turmoil of the French Revolution continued, a brash young soldier named Napoleon Bonaparte muscled his way to power. During Napoleon's turbulent reign, his consort Josephine carried on Marie Antoinette's tradition of blowing the national budget on expensive plants.

The irritable emperor frequently became furious at the immense sums of money the empress squandered on her botanical passion—millions of dollars at today's prices. Occasionally, however, Napoleon enjoyed a stroll in the relaxing gardens of Malmaison. He plotted some of his greatest political and military successes there, claiming that his ideas were loftier when out-of-doors.

Josephine's tempestuous husband frequently had to vie with plants for her attention. He was known to uproot her favorite plants when in a rage, which happened fairly frequently, as theirs was a stormy relationship, fraught with infidelities on both sides. But Josephine, if casual about marriage, was intensely devoted to her beloved garden. She ignored minor obstacles like France's long and bitter war with England, and obtained a special passport so that a London nurseryman could travel back and forth with new specimens for her ever-expanding gardens. She prevailed upon Napoleon to give orders to the admirals of the French Navy to be on the lookout, when capturing British ships, for crates or boxes bearing rare seeds or plants.

Josephine's garden, from roses to *Rhus*, was all about color. She may have consulted a best-selling garden manual of the time, James Meader's *The Planter's Guide, or Pleasure Gardener's Companion*, a guidebook for gardeners seeking to create dramatic masterpieces of color with leaves and branches instead of paint and canvas. It listed plants not only by family or species, but according to their place on the color spectrum, as though they were paints. Gardeners planned hues as carefully as an artist blending paints on a palette. By mixing and matching complementary leaf or flower shades, Josephine ensured that Malmaison would be an extravaganza of color, changing subtly through the seasons.

In her gardens, poison ivy rubbed shoulders with the most famous roses of the day. Row upon row of rose bushes as well as jasmine and hibiscus were backed by taller trees with dark foliage, against which ornamental vines like poison ivy and Virginia creeper would gleam brightly.

Josephine's marriage to Napoleon finally collapsed, largely due to her inability to produce an heir for the emperor. After receiving the shattering news of the impending divorce from Napoleon, Josephine sought refuge in her garden. It was autumn, and most of the roses were long gone, but she could find solace for her grief in the cheerful scarlet and gold of poison ivy.

—◆—

Poison ivy was a horticultural star for many years. But as Marie Antoinette and Empress Josephine discovered the hard way, fame is a fickle thing. As the eighteenth century waned and the nineteenth began, ornamental gardening's popularity as a hobby was spreading—not just among the wealthy, but among the common folk as well. As seeds and saplings from America and other exotic locations became more abundant and available, prices went down. Now everyone could afford poison ivy!

But sales began to decline. Because, of course, planting poison ivy could lead to unpleasant consequences. This wasn't an issue for queens and empresses, who weren't about to touch poison ivy with their own lily-white hands. Poison ivy could sustain its popularity with wealthy landowners, since they could afford large staffs of servants, and no one cared too much if the third assistant gardener broke out in a nasty rash. But as ornamental gardening became something that middle-class homeowners went out and did for themselves, the cultivation of poison ivy became understandably less popular. Poison ivy never became a beloved backyard friend like the daisies or petunias planted in cottage window boxes and village yards. French gardeners tended to refer to poison ivy not by its scientific Latin name, but by the less dignified moniker of *herbe de la puce*, or "flea grass."

As human enthusiasm for poison ivy waned, however, the plant began to grow in popularity with other species. Because there's a reason that poison ivy turns bright colors, and it has nothing to do with human needs or wants. The vivid coloration isn't intended to delight royalty, cheer up the lovelorn, or even to warn people off from its toxic touch.

Birds, especially berry-eating birds, generally have good color vision. They notice and are attracted to bright colors. The poison ivy plant needs to protect its seeds until they're fully matured and ready to create the next generation, so its ripening fruits are pale greenish-white, hidden under green leaves. But once the berries are ripe, the plant "wants" birds to come and eat them, so that the seeds can disperse. Poison ivy seeds are hard-shelled and relatively heavy, and they don't have parachutes or wings like many other wild seeds. But once a bird gobbles them down, the seeds can fly far and wide, encased in a coat of feathers.

American birds had evolved with poison ivy for millennia, but European wildlife didn't take long to catch on to the fact that the bright fall leaves signaled a bountiful crop of fruit. After feasting on berries in the royal gardens, a bird would fly over the palace wall and perch on a peasant's fence. And after the seeds had made the trip through the bird's digestive system, they germinated abundantly.

Like a bored princess slipping out of her castle, poison ivy escaped the confines of Malmaison, Versailles, and other aristocratic pleasure grounds and went slumming. The three-parted leaves began to crop up in hedgerows and ditches throughout England and Europe. Although beautiful as ever, poison ivy lost its cachet as a rare and exotic plant and was—and still is today—deemed an invasive nuisance. Something that sprang up in ditches and manure piles was no longer a fit candidate for fountained rose gardens. Poison ivy became a weed.

Spreading inexorably outward from the estates where it had been planted, it became a part of the landscape throughout Europe. For example, by 1840 it was reported as being widely naturalized in a botanical inventory of Malesherbes Park, on the outskirts of Paris. Today it's found on every continent except Antarctica—and flourishes in places as diverse as the elegant Kew Gardens in London and the hardscrabble Australian outback.

—◦—

After their separation, Napoleon allowed Josephine to retain the title of empress and keep the estate at Malmaison, where she lived for the rest of her life. Following his disastrous defeat at Waterloo, Napoleon retreated to Malmaison to ponder his fast-narrowing future options. His beloved Josephine had passed away a few years before, from a chill she caught while lingering in the garden on a cold day. The defeated emperor must have wandered, pensive and melancholy, through the rose gardens and under the poison ivy, regretting the past and contemplating his fate.

Alas for poison ivy, it too was on the verge of being ousted from royal estates and monarch's showplaces. Poison ivy's fifteen minutes of fame had lasted for more than two hundred years, but it finally came to an end—in the pleasure gardens of Europe, that is. But poison ivy was about to get some attention in its native land.

CHAPTER 8

A Virginia Native

Washington, DC, 1809

IT'S A TOUGH JOB TO BE PRESIDENT OF THE UNITED STATES. AND IT'S no secret that, after moving heaven and earth to get elected, more than one president has been relieved to come to the end of his final term. From George Washington to the present day, many chief executives have looked forward to leaving the cares of state behind and returning to private life. But of all the presidents anxious to leave the White House, none had his pants on fire more than Thomas Jefferson.

Abraham Lincoln hoped to travel when his last term ended. George Bush wanted to paint. But Jefferson was itching to get back to his garden. All through the years of the American Revolution, and the postwar political turmoil that followed, he had dreamed of Monticello, his mountaintop estate in Virginia. In the White House, he spent hours drawing plans and maps of flower beds, vegetable plots, and tree plantations. By the time he had finished two terms as president, he was desperate to get back to the soil he loved. The moment his eighth year in the White House was up, he fled. Three days after his successor was inaugurated, Jefferson was in the saddle, spurring his horse down the road to Monticello. He rode for hours through a March blizzard, so eager was he to get home in time for the planting season.

Jefferson wasn't planning to just kick back and relax in his garden. He was a man on a mission. He was eager to experiment with new varieties of plants—ornamentals, vegetables, crops, flowers. Most important of all, he was a patriotic gardener. He planned to prove that American plants were as good as—or better than—the plants of anywhere else in the world. And poison ivy was part of his plan.

We don't know exactly when Thomas Jefferson first encountered poison ivy in the wild—probably as an adventurous youngster exploring the Virginia countryside, then as now abundantly endowed with poison ivy. But he had to travel to Europe to discover poison ivy's potential as a cultivated plant.

In 1786, with the American Revolution only recently concluded, Jefferson and his colleague John Adams had visited Great Britain on an urgent diplomatic mission. Their assignment was to try to hash out some sort of commercial treaty with Britain, so that the struggling young nation of America could develop economically. But after months of negotiations, Adams and Jefferson had failed to make the slightest dent in the armor of British disdain. It was made abundantly clear to the two Americans that the former colonies were inferior in every way to Great Britain.

Fed up with slights and veiled insults, the would-be diplomats decided to flee London and soothe themselves with greenery. In early April they embarked on an ambitious garden tour. Visiting the gardens of the rich and famous had become a popular spring outing for Londoners, and the Americans joined hordes of English tourists. But Adams and Jefferson were more than casual vacationers. Lurching along in a carriage over muddy roads, they tirelessly covered hundreds of miles, sometimes checking out five gardens in a single day. They were eager to learn everything they could from the famous "nation of gardeners," for they were seeking to discover how to put together showplace estates of their own—American style.

And in every garden that Adams and Jefferson investigated, they made an astonishing discovery. America ruled. The only thing the British admired about the New World, apparently, was its flora.

American plants dominated British gardens. Jefferson might have thought he was back home in Virginia as he strolled through groves of white pine trees shading mountain laurel and rhododendron, while river birch, sweet gums, and sycamores had red springtime leaflets of poison ivy draped through their branches.

When Jefferson returned to the United States, he long remembered his garden tour. After his stint in the White House, he was eager to create a botanical masterpiece of his own. Not content with emulating English gardens, Thomas Jefferson planned to create a showplace that would outdo them all.

"Though I am an old man," Jefferson once famously remarked, "I am but a young gardener." As soon as he had made good his escape from the White House, Jefferson ordered thousands of seeds, saplings, bulbs, and roots. Teams of horses toiled up the winding road to his mountaintop garden, pulling wagons loaded with plants. The retired president happily spent the rest of his life among his beloved gardens, turning his Virginia acreage into a veritable Garden of Eden.

But, as with all Edens, there was evil lurking in the greenery.

Piles of volumes have been written about Jefferson's horticultural work, and one frequently encounters the phrase "Jefferson planted" this tree or "Jefferson planted" that flower. But the reality is a little different. Thomas Jefferson himself wasn't out there in the hot Virginia sun, digging holes, pushing wheelbarrows, and forking manure. He was out there in the garden directing his slaves where to plant things.

Monticello today is a meticulously re-created monument to one of the most famous of America's founding fathers. The gardens have been lovingly restored by volunteers as faithful replicas of the originals, containing the varieties of plants that were there in Jefferson's day, based on detailed research of his garden notes and inventories. But it's hard not to have mixed feelings while visiting this magnificent site. The eternal contradiction of this beautiful garden is that it was originally planted by enslaved hands. The amazingly liberal thinker who authored the Declaration of Independence never freed his own slaves. Faced with a choice between emancipation and financial ruin for his family (he was always deeply in the red and died tens of thousands of dollars in debt) he made a dreadful choice—one for which he realized history would condemn him. The beauty of Monticello, and the botanical research that Jefferson pioneered, would not have been possible without his workforce of slaves.

Jefferson's gardens were not only beautiful, they were revolutionary. Because almost half of the species planted at Monticello were native North American ones like poison ivy.

This was an unheard-of innovation. It was unusual in Jefferson's day and still is today. Most garden plants come from distant lands. Peonies originated in China, tulips came from Turkey, zinnias were wildflowers growing in Peru. Then and now, public parks and homeowners' yards are decorated with Norway spruces, Austrian pines, Siberian elms, Chinese chestnuts, and Japanese maples. But the fiercely patriotic Jefferson wanted to feature American plants.

Other Americans didn't share his enthusiasm for homegrown plants. It seemed to most gardeners that blossoms imported at vast expense from exotic climes were worth growing; the plants sprouting right at their own doorsteps or in the nearby woods and meadows were plainly weeds. But Jefferson stubbornly fought to eliminate this prejudice and demonstrate that New World plants were just as good as the tried-and-true Old World ones. Ironically, Jefferson had actually purchased many American specimens on a plant-shopping spree in London, where American seeds and bulbs were both cheaper and easier to obtain.

Jefferson deeply resented the "theory of degeneracy" espoused by French naturalist George-Louis Leclerc, comte de Buffon, which was all the rage in Europe. Buffon had written a highly regarded encyclopedia of natural history, *Histoire Naturelle*, which maintained that America's unhealthy, wet, and chilly climate caused all American species, both plant and animal—including humans—to become small, weak, and degenerate. Buffon pointed to Native Americans as an example of this unfortunate tendency. The irate Jefferson spent years debunking this theory, and eventually he presented Buffon with an enormous stuffed moose to prove that American wildlife was far from small and weak.

Jefferson was eager to prove that American plants were strong, too. He turned his acres of gardens into living laboratories, in which he experimented with both native and nonnative flowers, trees, ornamental shrubs, and vines. He kept detailed records, endlessly recording weather conditions and dates of plantings. He wanted to discover which plants were the most nutritious, which were the most beautiful, which were the likeliest to thrive in the Virginia climate.

Jefferson—or rather, his slaves—planted many species of vines. He experimented with grapevines, hoping to create home-brewed American wines. Many other species of vines, both native and nonnative, were planted for their ornamental beauty, including poison ivy. Jefferson put poison ivy in his list of ornamental plants, and *Rhus radicans* was included in his plans for Monticello landscaping, to be located in "the Open Ground on the West."

Thomas Jefferson and Marie Antoinette didn't have much in common politically, but they both favored the same style of naturalistic gardening. Poison ivy at Monticello, as at the Petit Trianon, was left pretty much in its wild state. This was as recommended in some of the many gardening textbooks that Jefferson consulted. A voluminous reader, one of his many inventions was a revolving book stand where he could keep five volumes open at once for easy reference. One of the books often spread wide was Phillip Miller's classic, *The Gardener's Dictionary*. In spite of the fact that Miller was an Englishman, his comprehensive book was Jefferson's bible when it came to planting methods. "Vines and other trailing plants," wrote Miller, "should be planted in Large Wilderness-quarters, near the Stems of great Trees, to which they should be trained up; where, by their wild Appearance, they will be agreeable."

This was a fairly reasonable way to garden with poison ivy. Training poison ivy vines to climb trees is about as difficult as training dogs to eat steak. Jefferson's enslaved gardeners had only to plant poison ivy rhizomes near a tree and stand back to let the plant do its thing. The young poison ivy vine reaches out, grabs onto the nearest tree trunk, and heads for the sun.

There are almost as many ways to climb trees as there are species of vines. Grapevines have curly tendrils that grab nearby branches. Virginia creepers use little suction cups. Asian bittersweet corkscrews its way round and round the tree trunk. But poison ivy's method of ascent is unique. Poison ivy gets to the top by growing thread-like rootlets all along the length of its stems. At first the rootlets appear as little patches that look like tufts of fuzzy hair. The furry roots secrete a glue-like substance that virtually cements the vine to tree or wall. As the main stem ages, more and more rootlets grow, forming a thick pelt of what looks like auburn fur—in fact, an old poison ivy vine looks more like a furry mammal than a plant. The wiry little roots dig their way into every tiny crevice in the bark and hold on tight.

Vines are parasitic: they derive benefits at the expense of a host organism. Most vines are harmful to the trees that support them, and some vines, like Asian bittersweet, can be lethal—they literally strangle their host, especially if the tree they're spiraling around is a young sapling. But with all those sticky rootlets clinging to the tree trunk, poison ivy doesn't need to corkscrew itself around the tree for support. It goes straight up. Poison ivy tends to prefer mature trees to climb, because the fuzzy roots can grab the older, rougher bark more easily, and so it's less likely to shade out young saplings. Poison ivy has minimal impact on the trees that host it, in fact, and is far less destructive a parasite than Asian bittersweet, grapevines, and many other vines. So Jefferson, who adored his trees, was not putting his beloved "pet trees" at risk when he decorated them with poison ivy.

Scratching the Itch: Deadly Smoke

If you tried to loosen a poison ivy vine's grip on a tree trunk (don't), you'd discover that it's so tightly welded to the tree that often the bark comes off with the vine. But the fuzzy vine, brown and inconspicuous, often goes unnoticed when trees are cut up for firewood. Logs with chunks of poison ivy vine grafted to their sides are sometimes thrown unnoticed into the fire. Tiny droplets of urushiol are carried by the smoke, and breathing it can be extremely hazardous to your lungs. Many people have suffered severe lung damage or even died from breathing urushiol-laden smoke.

Nowadays the re-created Monticello is almost eerily complete, down to the last detail. Jefferson's bedroom looks as though he has just left it, with his violin ready near his comfortable chair. His bed is made up with sheets and blankets, waiting for him to climb in. His revolving book stand is loaded up with garden books, ready to consult.

The gardens, too, are just as he planned them—with the notable exception of one species. Poison ivy is gone from the Monticello groves and gardens. Although it doesn't destroy the trees that host it, poison ivy has been eliminated to avoid irritating the tens of thousands of visitors who flock every year to see Jefferson's showplace.

But many other native species of plants still flourish in Monticello's woods and flower beds. Butterflies swarm to the red, shaggy blossoms of native bee balm. Monarchs find spots to lay their eggs on native milkweed. Goldfinches and sparrows feast on the seeds of native asters. Many of the native trees that Jefferson knew and loved are thriving. He would often drag reluctant guests out into the rain to view his favorite youngsters; now, two centuries later, the giant trees have fulfilled his vision of magnificence. Visitors to Monticello walk in the shade of hundreds of specimens of classic American species: tulip trees, white pine, and sugar maple. But the graceful vines of poison ivy that once decorated them are only a memory.

CHAPTER 9

The Vine Lifestyle

Or: Why Is Poison Ivy Such a Successful Plant?

To most of us, there's something shady about vines. Look at the names we give them—creepers, stranglers, parasites—they're the bad guys of the plant world. To those of us raised in a strict Puritan work ethic, there's something morally dubious about a plant that can't stand on its own two feet, so to speak. Why can't vines support themselves and not go draping themselves all over other plants?

A "clinging vine" is the very definition of weakness. But the great naturalist Charles Darwin, in his groundbreaking research into the mysteries of evolution, concluded that vines are among the most powerful and highly evolved plants on the planet. Vines like poison ivy are extremely successful plants, able to thrive in places where more upright, self-supporting types can't. This is because vines, while remaining firmly rooted to the earth, can move in amazing ways that other plants haven't mastered. They are, in fact, eerily non-plantlike, displaying an awareness of their surroundings and heading toward their goal with an almost human purposefulness.

Darwin was endlessly intrigued by vines. He spent decades studying and experimenting with more than a hundred varieties of climbing plants, including poison ivy. And what enthralled him the most was this uncanny power of movement.

After Darwin had returned from his famous voyage around the world, he settled into a quiet rural life, in a pleasant manor called Down

House set in the verdant countryside south of London. Green vines draped the white plaster of the walls, and perhaps they reminded him of tropical vines he had seen on his travels. In any case, while pondering evolution and the origin of species, he spent years studying vines. His methods of research were described in a book titled *The Movements and Habits of Climbing Plants*.

Often subject to bouts of a mysterious, debilitating illness, Darwin spent many hours in bed. But never one to waste time, he observed vines in pots on his bedside table. He began his study by observing a small hop vine. "To ascertain more precisely what amount of movement each [vine] underwent," he began, "I kept a potted plant, during the night and day, in a well-warmed room to which I was confined by illness." He timed, charted, and tracked the plant's every move.

Now, if you've ever spent an afternoon watching a plant grow, you'll know that it can be rather a dull way to pass the time. But Charles Darwin was nothing if not patient. Lying there hour after hour, he observed that the tip of the young vine slowly bent itself at an angle. Then, groping like a blind man's hand, the vine rotated to all points of the compass, turning a full 360 degrees in a twenty-four-hour period.

Darwin, intrigued, got up from his sickbed and put a stick where the vine would come across it. He watched as the plant reached the stick and began to twine around it, spiraling steadily upward. The vine seemed to be *looking* for something.

Plants don't have eyes, of course. But they have cells called photoreceptors, which, in Darwin's terms, are "excited" by light. Vines actively seek light out, like a hungry animal foraging for food.

Because to a plant, light *is* food. Photosynthesis is the name we give to the complex chain of chemical reactions that occurs in every green leaf: the miracle of turning air and water into sugars that nourish the plant. Sunlight is the start to this process, the key that turns the engine, so to speak. Most plants stay put and wait around until a ray of sun finds them. But poison ivy goes looking for excitement, seeking a path to ever more light.

Of course the whole plant doesn't get up and walk around. A vine remains rooted to the earth in at least one spot, but it can cover a lot of ground. It can even leap tall buildings.

A vine seed germinates in a pocket of soil, say, in a narrow crack between a sidewalk and a wall. The young vine creeps upward, spreading out over potential acres of a habitat—the vertical brick wall—that almost no other plant can use. This trick works on all sorts of buildings, like elite colleges (they don't call it the *ivy* league for nothing) as well as cottages, cabins, and castles, and of course in more natural habitats like steep cliffs or boulders.

Among all the thousands of species of vines, poison ivy excels at this strategy of exploiting a vast range of habitats. It can grow on the side of a rock-strewn mountain, or climb a telephone pole, or shade a desert cliff or a construction site. I once attended a wedding at which the bride and groom posed for pictures in blissful ignorance under a trellis wreathed in poison ivy.

Poison ivy can survive in a great variety of light levels, from full sun to deep shade. It has a high degree of a quality that botanists call "physiological plasticity," meaning it can adapt its shape and structure, and even its internal processes, to differing conditions. Each individual leaf can change subtly, acclimating itself to the light conditions in which it

Scratching the Itch: Take a Cold Shower

If Charles Darwin chanced to develop a rash in the course of his poison ivy research, he didn't record it. But it's likely he would have turned for a cure to his favorite panacea: hydropathy, or the water cure. Darwin suffered for much of his life from various ailments that were never fully diagnosed. He was convinced that frequent cold water baths and showers were good for his health. In the case of an encounter with poison ivy, he was right. Abundant application of cold water is recommended for washing away the urushiol after contact with the plant. And if a rash develops, compresses or rinses of pure cold water can help soothe the itch without damaging sensitive skin.

finds itself. One reason that Darwin was so interested in vines was that their incredible variety and complexity offered an opportunity to prove his theories of evolution and natural selection. Darwin considered vines to be the most highly evolved of all forms of plant life.

"Vines have wonderfully little expenditure of organized matter," Darwin noted, "in comparison with trees, which have to support a load of heavy branches by a massive trunk."

In a way, a tree's strength is its weakness. A tree, of course, has a strong trunk, to hold up all those leaves to the sun. Therefore the tree has to have lots of leaves, to do lots of photosynthesis to create those thick-walled cells that make up the trunk. As the years pass, the tree gets taller and wider, each inch of growth requiring more strong cells to hold up the leaves to the light so they can make more food . . . and so on. Trees start out as bendable saplings in their youth. But trunks get more rigid as they age, till they finally become too brittle to resist the forces of gravity or wind.

A vine, on the other hand, has flexibility at its core. The apparently weak stems are actually as versatile and strong as wire. Vines like poison ivy that attach themselves firmly to a trunk don't need strong, stiff cells to hold themselves up. Even a six-inch-thick poison ivy stem is so bendable you can tie knots in it. (Don't try this at home.)

And even if (when) the tree falls down, the poison ivy vine that envelops it will bend but not break. If its host collapses, the vine just gets busy creeping about until it encounters another host to climb. Poison ivy vines can actually become *less* rigid with age. There's a life lesson there somewhere.

———

Charles Darwin was a diligent and disciplined worker. And he was a little scornful of the types of vines that merely leaned lazily against a surface. He considered that vines that actively climbed using tendrils or rootlets, like poison ivy, were more highly evolved organisms.

Still, when you see a vine draped over an unfortunate tree, it's hard not to think of the vine as a shameless freeloader. But there is no such thing as a free lunch. The vine lifestyle is a very successful one, except for one problem. A vine is essentially a long, thin tube, stretching for perhaps hundreds of feet. A human-like network of veins stretches through it, transporting sap and nutrients from leaf bud to root tip and back again. But like the human vascular system, it can break down if overstressed.

An embolism is a gas bubble or a blockage that disrupts the flow of sap, which can cause the plant to (so to speak) have a heart attack and die. This limits the plant's possibilities. A vine can't grow to an unlimited length.

One of the major causes of embolism in vines is the freezing and thawing that occurs in temperate climates. This is why so many species of vines grow in tropical latitudes. Winters are tough, as we all know, and not many vine species can cope with the severe physiological stress of extreme cold. But poison ivy can.

Poison ivy vines are superbly adapted to keep the sap flowing efficiently in severe conditions: heat or cold. The veins are more like a multistranded cable than a single hose. The vine can be twisted and bent, but the sap inside continues to flow in an unbroken stream. Again, it's all about being flexible.

Darwin did most of his work with vines toward the end of his life. Weary of the endless controversy his groundbreaking ideas on humans and apes and other animals had stirred up, he spent more and more time pottering around with the soothing greenery of plants. "It has always pleased me to exalt plants in the scale of organised beings," he wrote.

The last portraits and photographs of Charles Darwin, taken in his old age, show a man bent and white-bearded. Most of us, if we're lucky enough to reach the age at which our hair turns white, get set in our ways, stiff and inflexible as a tree trunk. As did Darwin, apparently. For decades he lived a life of scheduled predictability at Down House, breaking his days into sections for eating, reading, research, letter writing, etc. He took the same walk every day at the same time, and played two games of backgammon with his wife Emma every night for decades.

Which is not to say that his thinking became inelastic. In Darwin's photos he seems humble, almost apologetic, as though he's sorry to have wrought such an enormous upheaval on the world with his shattering ideas about science and evolution. Nothing in the appearance of this old man shows the amazing flexibility of his mind. It seemed that the older he got, the less rigid his thinking became. His genius developed as he matured, giving him the strength to turn the world upside down. A very successful maturity—as versatile and flexible as a poison ivy vine.

CHAPTER 10

The Columbian Exchange

Jamestown, Virginia, 2015

THE BEARDED MAN STANDING IN FRONT OF ME ON THIS BRIGHT SPRING morning could be Captain John Smith himself. He wears a starched ruff and a jaunty hat with a feather, and I can see my face reflected in his polished steel breastplate. Every detail of his clothing, his weapons, and his appearance is a painstaking attempt to re-create the past. But there's one tiny clue that blows his cover.

With a glance, I can tell that no miracle of time travel has occurred. I have definitely not been transported back to 1607. Captain Smith's booted feet are solidly planted on a lawn of white clover (*Trifolium repens*)—which wasn't imported to the New World until the eighteenth century.

———

Historians generally do a bang-up job of re-creating all the details of times long gone. The sights and sounds, even the smells of history are brought to life at living history museums like Historic Jamestowne, a re-creation of the first permanent settlement in America. You can sniff the scents of wood smoke and authentic horse manure. Students on field trips thrill to the bang and spark of muskets being fired. Perspiring reenactors (it's *hot* in Virginia) in authentic wool clothing churn butter or hoe tobacco with period tools. Here are sounds not heard for centuries, brought back to life: the scrape of clamshells as Native American interpreters hollow a dugout canoe, for instance. Agricultural historians even back-breed cows and sheep, painstakingly tracing genetic lines, selectively breeding crops and farm animals to produce smaller, scrawnier versions that the early colonists would recognize. A really good living history museum is the next best thing to time travel.

But as far as wild plants go, the historians flunk out big-time. Historically accurate forests and meadows are in short supply. The grass underfoot and the trees overhead have changed dramatically in the relatively short time span of four hundred years. When the reenactor portraying John Smith glances around, hardly a species his gaze falls on could have been seen by the real John Smith in 1607.

What happened? The land hasn't all been paved. There's plenty of greenery around here—the Jamestown site is surrounded by trees. Woods is woods, right? Obviously if you cut the trees and build a shopping mall, the landscape changes. But here plants are, all around us, green and growing. What's so different?

~

Beginning in the late fifteenth century, a change occurred on our planet. It was a huge, ecosystem-shattering change. Think of the asteroid that hit the Earth and wiped out the dinosaurs—like that. But this change was far less dramatic than a monster meteor slamming into a Tyrannosaurus rex. This change was so subtle that only a tiny percentage of the human population even noticed it was happening. It's that green blindness thing again.

Until the fifteenth century, the New and Old Worlds had been more or less biologically separated for millions of years. Perchance a few hardy Vikings or Irish monks had reached the Americas, but their effect on the environment was minimal. But then came the year that all American schoolchildren commit to memory—1492. When Christopher Columbus took his first steps onto that New World beach, everything began to change.

There weren't only humans on board the *Niña*, the *Pinta*, and the *Santa Maria*. A few dozen Norway rats probably set foot on shore before Columbus did. Houseflies. Fleas. Lice. And plants, of course. If you've ever dug a barbed burdock out of your sock or picked a hitchhiking seed off your pant cuff, you know how good seeds are at stealing rides. Perhaps there was a seed in the crack of Columbus's boot sole, a burr stuck to the hem of his cloak. Living organisms were beginning to move from the Old World to the New.

And of course the flow went both ways. Columbus promptly loaded up his ship with curiosities from the new lands he was so proud to have "discovered," never giving the rights of the "unchristian" inhabitants a thought. He brought back many biological souvenirs: pineapples, cotton, tobacco, turkeys—and whatever insects or parasites were attached to them. The barriers of the Atlantic and Pacific Oceans were shrinking fast. Bit by bit, with every ship that sailed to the New World, germs and insects and seeds came, too. Ecologist Alfred Crosby was the first to give this phenomenon a name: the Columbian Exchange.

At first, practicality determined which organisms were purposely transported. The colonists had to eat, and the wilderness didn't seem to offer much in the way of livestock or crops. The settlers brought sheep, cows, and pigs as well as the crop plants they were familiar with (wheat, oats, barley) and garden vegetables (carrots, peas, turnips).

But surprisingly, they found that the Americas did indeed have a few kinds of plants you could eat. Some of them were pretty darn good, too. Like chocolate, for instance. And pecans. Plants from the New World could be nutritious, or intoxicating, or delicious. Potatoes, pumpkins, tomatoes, corn, tobacco, vanilla, peppers: all came from the New to the Old World.

This two-way street was also true of medicinal plants. Anxious mothers, concerned about the dangers of the New World, brought their own remedies, often in the form of seeds for the herb garden. Plants like wormwood, feverfew, woundwort, sage, mint, and dandelion were the all-night pharmacy of the colonists. But of course the Europeans soon discovered a wealth of American medicinal plants. Exporters grew rich shipping medicinal plants like sassafras, ginseng, and goldenrod back to the Old World.

When the real John Smith and his fellow colonists arrived at the real Jamestown in 1607, they immediately began cutting down trees to build a palisade for protection from the Indians. They were undoubtedly too busy to notice the weeds that sprang up in the new sunlit clearings, in the disturbed soil around the fort, in their gardens and latrine pits. Nature abhors a vacuum. Every time a tree was cut, a poison ivy vine ripped out, a new plant took its place. And often it wasn't a New World plant that sprouted up, but an Old World one.

Peering out through the shadows, keeping a close eye on the bizarre newcomers, the Virginia Indians began to notice the changes. One plant, common plantain, was so ubiquitous that Indians began referring to it as "Englishman's footsteps" since it had an oval leaf that resembled a booted footprint. Honeybees, imported by the colonists to pollinate their orchards, were referred to as "English flies." Everywhere the newcomers went, everything changed—even the insects buzzing past and the plants underfoot.

White and red clover, daisies, daylilies, Queen Anne's lace, chicory: these now-common roadside weeds evolved in the sunny climes of Asia or the Middle East, lovers of sun, tolerant of plowing. These hardy plants had spread through the Old World since the dawn of agriculture, and now they exploded into the new territory opened to them. By 1672, when the first systematic botanical survey was carried out in New England, dozens of species of European plants had become a common sight around farms and villages.

Historian Charles Mann, citing Crosby, argues that "the Columbian Exchange underlies much of the history we learn in the classroom—it was like an invisible wave, sweeping along kings and queens, peasants and priests, all unknowing." Think about it—it's a little unsettling. Horses, smallpox, sugar, rubber, coffee, tea, daffodils, influenza, potatoes, leprosy, pigs, and poison ivy: in a relatively short space of time—only a few centuries—thousands, perhaps millions of species moved to new continents. The planet would never be the same.

John Bartram, the Quaker farmer who shipped his famous seed boxes all over Europe, was a significant contributor to the Columbian Exchange. He was proud to be personally responsible for approximately a quarter of all the American plants sent to Europe during the colonial period. In return for the tens of thousands of plants and seeds he dispatched overseas, Bartram received baskets and crates of exotic plants, both useful and beautiful, as gifts from his grateful customers: Persian cyclamen, Swiss barley, Chinese rice, Scotch cabbage, and a host of other species from all quarters of the globe.

Bartram eagerly planted many of these specimens in his own garden and also sold cuttings and seeds to American gardeners. Seeds of the beautiful Norway maple tree, for example, were sent to him by a European admirer, and Bartram propagated them and sold many thousands of these scenic, sturdy shade trees to American homeowners. In his enormous garden, a peaceful spot on the banks of the Schuylkill River, a mix of American, Asian, and European species intermingled, Virginia creeper and poison ivy twining through Norway maples and Chinese ginkgoes.

One summer day in 1759, Bartram noticed that a pretty little flower was spreading through his garden. Toadflax, it was called, a dainty yellow and orange blossom, affectionately called "butter-and-eggs" by children. It had been sent to him as a gift by a friend from England, and Bartram had planted a small patch. But despite constant weeding, toadflax had crept into other beds as well. In fact, Bartram noticed, with a small feeling of uneasiness, the pretty little flower was proving quite impossible to get rid of.

Bartram saw the humble toadflax flower as a warning bell. He mentioned the persistent weed in a letter to his British colleagues, complaining of "plants that are most troublesome in our pastures and fields in Pennsylvania; most of which were brought from Europe." He listed plants like dandelion, ox-eye daisy, scotch thistle, mullein, and dock, species that "have escaped out of our gardens and taken possession of our fields and meadows, very much to our detriment," titling the list "The Worst Plants Introduced to the New World."

A bit late in the day, John Bartram was beginning to realize that transporting plants all around the globe might not be the great idea it had seemed to be. He was a self-taught man, not a classically trained scholar, so he may not have thought to compare his famous boxes to the ancient Greek legend of Pandora's box, one of humanity's oldest myths.

Pandora was given a container of marvels by the gods and told never to open it. But she wasn't a goddess, only a human, and so she couldn't resist. Eager to see all the wonderful things inside, she opened the lid. The things that flew out were soon beyond her control. Bartram was beginning to realize that his boxes were, in a sense, very like Pandora's. Once the lid was opened, the contents could never be put back inside.

John Bartram had discovered a simple and powerful truth: when you mess around with nature, bad things can happen. But his list of "Worst Plants" was ignored, and his warnings fell on deaf ears.

It seemed so natural, so right, to collect useful and beautiful plants and share their blessings with others. What on earth could be wrong with introducing a species of plant to a new place, if the plant could be useful—if it could feed hungry people, prevent soil erosion, cure disease, or soothe the soul with its beauty?

Plants have wonderful benefits and beauties, it's true, but we forget that plants are shockingly aggressive organisms. Unable to move out of our way, vulnerable to the lawnmower or the chainsaw, plants seem like passive organisms—alive, yes, but hardly capable of putting up a fight. But plants can be brutal. Plants beat back their predators with spikes and thorns, poison their competitors, strangle their hosts. They just do it so quietly that we don't notice. Green blindness.

Bartram brought toadflax to the New World and sent poison ivy to the Old, but what he forgot to add to his boxes were the natural mechanisms that could keep these plants in check. A complex combination of predators, pollinators, fungi, and bacteria have coevolved with each and every plant. There are patterns of weather, temperature, shade, and countless other factors that the plant has to cope with in its native habitat. But each time Bartram carefully uprooted a specimen or packaged a seed or bulb, he left behind the plant's natural restraints.

Some of the species Bartram introduced to America are now considered invasive disasters. Pretty little toadflax is no problem at all compared to the Norway maple, which is quite possibly one of the half dozen most destructive plant species on the continent.

What!? How can such a pleasant, graceful shade tree be destructive? Norway maple is indeed a lovely tree, with attractive foliage, and very easy to grow. Its thick canopy of leaves provides delightful shade in summer—an obvious choice for planting in the backyard, as Bartram thought.

But the heavy shade cast by a Norway maple kills every native plant that grows near it. The tree's fast-growing, shallow roots create a thick web that chokes out competitors. Once a Norway maple invades a forest, it's good-bye to the ferns, orchids, trilliums, and lady's slippers as well as seedlings of native trees like oak, beech, and sugar maple. Norway maples produce prolific crops of seeds shaped like little helicopter blades that whirl on wind currents far and wide, damaging forests across America.

Or then—to choose an example from my own backyard—there's a pretty little wildflower called garlic mustard. Sounds tasty, doesn't it? Garlic mustard is a biennial plant in the mustard family, an attractive Eurasian wildflower, and as the name implies, it is delicious indeed. Every part of the plant, the dainty white blossoms and the scalloped leaves, have a savory taste of garlic. Imported to this country as a condiment in the mid-1800s, it soon became a popular garden herb, as it's easy to grow just about anywhere—in fact, it doesn't even need a lot of sun. Garlic mustard grows really, really well in the shade. And that's the cause of the catastrophe unfolding in my yard and countless others across the country.

Shade is the barrier most plants won't cross. Therefore most nonnative flowers don't pose a threat to wilderness areas. Tulips, petunias, peonies—they're none of them native American plants, but they stay where they're put. Even the hardy nonnative "weeds" like dandelions, clovers, daisies, daylilies—they all stick to the open fields and roadsides and don't venture onto the backcountry trails. They need sun and can't survive shade.

But shade-tolerant garlic mustard can march across a forest like storm troopers, eradicating the native vegetation. No pink lady's slipper can stand against the jackboots of garlic mustard. As ferns, grasses, moss, and wildflowers disappear under the scalloped leaves of garlic mustard, thousands of acres are being fundamentally altered by this innocent invader.

It's not that garlic mustard or Norway maples are the evil Nazis of the plant world. No plant is inherently bad. It's just lacking the natural restraints that should keep it in check. On the North American continent, almost nothing eats garlic mustard—bugs ignore it, even white-tailed deer rarely nibble it. Diseases and fungus it might have had to battle in its homeland aren't present here. Safe from predation, it outcompetes almost everything—although a high-climbing poison ivy vine is one of the few forest plants that can rise above a garlic mustard infestation.

~~~

Plants are best left in their native habitats. Like most of nature's lessons, this one is taking us a long time to learn—the concept of being wary of invasive plants has taken centuries to sink in. When I first bought a house with a yard and had a garden of my own, I wanted to plant species that would be good for wildlife, especially birds. But in those pre-Google days, who to ask for advice? I consulted garden centers and plant nurseries, called my state wildlife agency, and perused publications put out by the federal government, trying to discover which shrubs, trees, or vines would be best to plant.

And no matter where I inquired, the answers were the same. You want to plant Russian olive, I was told. Great for birds. Definitely get some Tatarian bush honeysuckle, an easy-to-grow Eurasian shrub. Multiflora rose, a hardy native of China—birds love it! And absolutely, I was assured, plant Asian bittersweet. A 1979 publication of the US Department of Agriculture urged responsible homeowners to plant multiflora rose, Manchurian crabapple, and Russian olive for erosion control. My state wildlife agency even gave away "wildlife shrub packets" of bush honeysuckle and multiflora rose for free! Seemed too good to be true.

No one, not one source, suggested planting poison ivy, for some strange reason. But what's even stranger is that almost no one recommended planting *any* native species at all.

Obediently, I planted all of those nonnative species in my yard and out in the back acres. Who would turn down a free bundle of "shrubs for wildlife habitat"? But as I'm sure you've discovered, if something seems too good to be true, it usually is. I've spent a good deal of the last twenty years trying to eradicate them.

Asian bittersweet does indeed have berries that are attractive, to humans because of their lovely orange hue, and to birds as well. But bittersweet vines are lethal to forests. The prolific vines corkscrew up tree trunks, overshading the branches and eventually strangling the trees. A woodlot infested with bittersweet vines turns in short order to a thick shroud of bittersweet draped over the skeletons of dead and dying trees.

Russian olive and multiflora rose stick to open, sunny ground, where they spread thickly, taking over whole meadows. They choke out native wildflowers and fruit-bearing shrubs and finally render the meadow impassable to anything but a Sherman tank.

But I've fought my hardest battles against the highly invasive honeysuckle. Seems like honeysuckle would be a good thing—but it's not the same plant as the fragrant vine that grows down south. Bush honeysuckles were imported from Japan, China, Korea, Manchuria, Turkey, and southern Russia. The little bushes with pink and white flowers look pretty and innocuous, but the shaggy gray stems are as resistant to clippers as wire cables, and the bush resprouts with enthusiasm after being mowed. It's harder to eliminate from a forest than bedbugs from a mattress.

And so it continues to this day. Walk into any tree nursery and try to find a native American tree for sale among the Japanese maples and the Serbian spruces. Norway maples are still for sale across the nation, a wildly popular backyard landscaping choice. True, people are becoming more aware of the issue, and many nurseries are valiantly trying to include native plants among their wares. But it's so much easier to grab a flat of petunias or marigolds (native to South America) or stick a few tulip bulbs (native to Turkey) in the flower bed.

Customers like plants that are easy to grow and pest resistant. And nonnative species fill the bill. A native plant—a red oak tree, for example—can potentially be nibbled and chewed by hundreds of species of insects, thereby making it a restaurant for bug-eating birds. A nonnative species like Norway maple or Chinese ginkgo will have unchewed leaves—but the bugs and the birds go hungry. And if you doubt the link between insects and birds, just page through Audubon's *Birds of America*—he drew almost as many insects and spiders as he did birds. In fact, there's a bug or a bug-chewed leaf on almost every page—sometimes the hapless insect is portrayed still kicking in the bird's beak.

It requires a revolutionary gardener to welcome munching caterpillars, leaf miners, and hungry beetles into the garden. It's hard to turn your head and look away from holes in the leaves you've so carefully nurtured. But that's the way nature works.

When Thomas Jefferson insisted on planting native species in his garden, he was right—although for the wrong reason. He was making a political statement, not an ecological one. As a patriot, he was trying to show that Yankee plants were just as damn good as European ones.

So they are—but only in America. The native American bee balm he planted attracts ruby-throated hummingbirds—but ruby-throats aren't found in Europe. European birds need different plants. Jefferson's native butterfly weed is a crucial host plant for American butterflies. But English butterflies have different host plants.

We're only now beginning to pay attention to this complex lesson. For it's not only a question of aesthetics—whether we can enjoy the pretty sight of a pink lady's slipper or a lacy fern. There is much at stake here, as our landscape slowly, subtly changes. Because plants are the essential building blocks of habitat. Native plants like poison ivy are the roots from which spring all of America's wildlife.

The first spring leaflets of poison ivy. DOUG WECHSLER

Captain John Smith, the first person to describe in writing the "rednesse, itchynge, and lastly blisters." HOUGHTON LIBRARY, HARVARD UNIVERSITY

Audubon print. UNIVERSITY OF PITTSBURGH

Like John Bartram, poison ivy climbs high.
DENISE HACKERT-STONER, NATURELOGUES

Linnaeus's experimental garden at the University of Uppsala in Sweden, where he enthusiastically cultivated poison ivy as well as many other North American plants.
ANITA SANCHEZ

Dr. Dufresnoy wondered if poison ivy had the potential to be a hero.
DENISE HACKERT-STONER, NATURELOGUES

Seeds of poison ivy (lower right) and pokeweed removed from a single dropping of a young catbird. JULIE CRAVES, ROUGE RIVER BIRD OBSERVATORY

The manicured gardens of Versailles, where monarchs intended to demonstrate their mastery over nature—although poison ivy is a hard plant to master. ANITA SANCHEZ

At Monticello, Thomas Jefferson championed native plants like poison ivy and bee balm, eager to prove that American plants were just as magnificent as Old World species. ANITA SANCHEZ

Charles Darwin experimented with poison ivy and considered it one of the most highly evolved plants on the planet. LIBRARY OF CONGRESS

Hairy vine, a danger sign!
GEORGE STEELE

Yellow-rumped warbler feasting on poison ivy berries.
JIM MCCORMAC

Poetic justice: a mite infestation makes poison ivy leaves look as though they have
a terrible rash. BILL BAILEY

The qualities that make poison ivy so hard to eradicate make it priceless when it comes to protecting the soil. Eastern Point Lighthouse, Gloucester, MA.
ANITA SANCHEZ

The toxic yet healing touch of poison oak.
JUDITH KRAMER

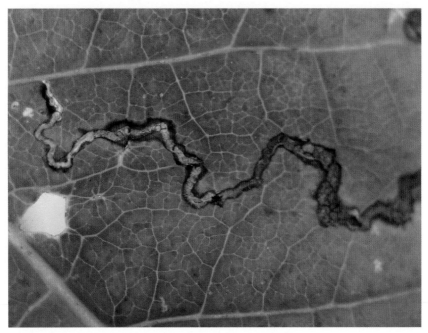

A leaf miner insect chews a trail through a poison ivy leaf.
ANITA SANCHEZ

A bee sipping poison oak nectar. Herbicides are taking a deadly toll on the world's honeybees and other crucial pollinators. JUDITH KRAMER

Poison ivy ascending a white pine.
MICHAEL LEE

Poison ivy takes a bow. ANITA SANCHEZ

Can you spot the poison ivy?
ANITA SANCHEZ

Berries white, take flight!
FRANK KNIGHT

Poison ivy's infinite variety—like snowflakes, no two alike.
DENISE HACKERT-STONER, NATURELOGUES

CHAPTER 11

What Do Animals Eat?

Ossining, New York, 1896

THE WHITE-COATED SCIENTIST LAID THE TINY CORPSE GENTLY ON THE laboratory table. He parted the fluffy feathers on the breast. With a knife blade honed to razor sharpness, he made a delicate incision into the tissue-paper skin. Opening up the sparrow's stomach, he carefully removed the contents.

Squinting at the tiny fragments, he examined them to determine what each speck might be. A half-chewed seed? A fragment of insect wing? He recorded the data with painstaking exactness:

Scarabaeid [beetle] remains 10%
Elytra [parts of insect wings] of Chrysomelid? [leaf beetle] 10%
20 seeds of Dactylis glomerata [orchard grass] 70%
2 Rubus [blackberry] seeds 10%

Then he signed his name: *Sylvester D. Judd*. He labeled the card *#1* and filed it carefully away.

◄━━►

Back in the days of John Smith, John Bartram, or Thomas Jefferson, it seemed as though America's bounty was limitless. Mighty oaks, towering pines, eagles, salmon, beavers, moose, bison . . . there was apparently no end to the riches of the New World. And so there were no laws protecting that which was so abundant—anything was, quite literally, fair game. If a hawk was threatening your chickens, well, just shoot the hawk out of the sky. Want a snowy egret's plume to decorate your hat? Go right ahead. Quail stealing your grain? Just eliminate the quail. Why not? Plenty more where that came from.

Well into the nineteenth century, there were only a few individuals who realized that nature was not inexhaustible. Wild plants and creatures were admired for their beauty, or cursed as a nuisance, depending on your point of view. But few people realized that there might come a day when they wouldn't be there at all.

Dr. Sylvester Judd was one of a handful of pioneers in a field of study that was just taking shape in at the end of the 1800s: wildlife management. His research began as an attempt by the government to find out what kinds of pesky birds were eating farmers' grain crops, but over the years it evolved into something much more complex. Perhaps Judd himself didn't realize how valuable his research would ultimately prove to be. We can only be grateful that he did it. Because the little card with information on the sparrow's meal was only the first in a very, very long line of similar cards filled out by him and other scientists—a quarter of a million nuggets of information about the eating habits of American wildlife.

Scientists were beginning to realize that survival for the nation's wildlife lay in preserving the habitat that sustained it—especially the key plants that provided food. But finding out what wild animals eat can be a challenge. You could try to watch an animal at its meal, yes. But it's almost impossible to tell from observing a chewing moose or a flitting chickadee exactly what the creature is eating, and exactly how much. However, there's one sure way to discover precisely what an animal has ingested, and it's not for the fainthearted.

These unsung heroes of the early days of wildlife research dissected huge piles of moose and bison droppings. They inspected tiny specks of mouse scat, and the cigarette-shaped excrement of wild turkeys. They investigated the contents of the cheek pouches of pocket gophers, slit open the crops of hummingbirds and woodpeckers, opened the gizzards and stomachs of hundreds of thousands of mice, rabbits, deer, snakes, lizards, and fish. They spent thousands of hours patiently piecing together bits of seed hulls, identifying fruit pits, and assembling insect fragments and rodent skeletons like insanely small jigsaw puzzles.

Day after day, year after year, these patient scientists recorded precisely what each animal had consumed, and in what proportions. This rich horde of data, slowly, painstakingly amassed for more than half a century, is preserved on yellowing index cards—thousands of them, neatly numbered—in the archives of the Fish and Wildlife Service in Washington, DC.

Sometimes (despite much evidence to the contrary) a government agency does something right. The United States Fish and Wildlife Service was created in 1940, but had its origins as the US Commission on Fish and Fisheries in the Department of Commerce in 1871. Through the years it employed many brilliant scientists (one of its few female employees, a marine biologist named Rachel Carson, eventually went on to write a shocking best seller about pesticides). After half a century of data on wildlife foods had been amassed, three biologists, Alexander Martin, Arnold Nelson, and Herbert Zim, were commissioned to put it all together.

In that long-ago era before the personal computer, it took a while to coordinate this enormous mass of information. They crunched the numbers, added up the percentages, and finally put all that wealth of observation and knowledge into a five-hundred-page volume. *American Wildlife and Plants: A Guide to Wildlife Food Habits* became the wildlife biologist's bible.

The accumulated research yielded some very surprising results. Farmers had long assumed that wild birds and mammals were nothing but crop-eating pests, but the *Guide* proved conclusively that birds ate enormous numbers of caterpillars, wireworms, beetles, and other insects that despoiled grain. Foxes had been widely considered chicken-eating vermin that hunters were encouraged to shoot on sight—but the *Guide* revealed the amazing numbers of rodents eaten by wild canids like foxes and coyotes.

There was an immense amount of practical information—something for everyone. No longer did hunters have to guess if one of the reasons for the decline of the wild turkey was a lack of acorns. Or bird-lovers wonder what foods cardinals need to make it through the winter. Want to know what wildflowers are most attractive to the ruby-throated hummingbird? Turn to page 116. Curious about the dietary habits of the thirteen-lined ground squirrel? Check out page 249.

The book resoundingly answered the question, what do animals eat? And for a remarkable number of species, the answer was poison ivy.

—◦—

Poison ivy's small, hard seeds are quite easy to identify. They're distinctively rounded, almost heart-shaped. They were found in the droppings or innards of more than fifty species of birds. Harder to identify were the chewed remnants of poison ivy leaves, but direct observation confirms that they are an important browse for many species of mammals.

All those decades of research, as well as research ongoing to this day, demonstrate that poison ivy is a major wildlife food. In the *Guide*'s listing of hundreds of native woody plants that are of survival value to wildlife across the nation, poison ivy was ranked seventh.

The work continues. In the fall of 2011, the blog of the Rouge River Bird Observatory, run by the University of Michigan–Dearborn, reported that researchers had discovered the seeds of poison ivy in the droppings

Species That Feed on Poison Ivy and Oak

From *American Wildlife and Plants: A Guide to Wildlife Food Habits*, by Alexander C. Martin, Herbert S. Zim, and Arnold L. Nelson, published 1951.

The book considers poison ivy (*Toxicodendron radicans*) and poison oak (then known as *Toxicodendron quercifolium* and *Toxicodendron diversilobum*) together. At that time poison ivy was considered to be one species; nowadays it's divided, like poison oak, into two species.

Animal nomenclature has also changed considerably over time.

Birds

Sharp-tailed grouse
Ring-necked pheasant
Bobwhite quail
California quail
Wild turkey
Eastern bluebird
Bush-tit
Catbird
Black-capped chickadee
Carolina chickadee
Chestnut-backed chickadee
Mountain chickadee
Crow
Purple finch
Red-shafted flicker
Yellow-shafted flicker
Oregon junco
Slate-colored junco
Ruby-crowned kinglet
American magpie
Yellow-billed magpie
Mockingbird
Phoebe
Red-breasted sapsucker
Yellow-bellied sapsucker
Townsend solitaire
Fox sparrow
Golden-crowned sparrow

White-crowned sparrow
White-throated sparrow
Starling
Brown thrasher
California thrasher
Hermit thrush
Russet-backed thrush
Varied thrush
Plain titmouse
Tufted titmouse
Spotted towhee
Warbling vireo
White-eyed vireo
Audubon warbler
Cape May warbler
Myrtle warbler
Cedar waxwing
Downy woodpecker
Hairy woodpecker
Lewis woodpecker
Pileated woodpecker
Nuttall's woodpecker
Red-bellied woodpecker
Red-cockaded woodpecker
Cactus wren
Carolina wren
Wren-tit

Mammals

Black bear
Mule deer
Pocket mouse

Muskrat
Cottontail rabbit
Wood rat

This listing has been added to by many other biologists over the years. For instance, both moose and white-tailed deer browse on poison ivy, as do many other species.

of several bird species, particularly woodpeckers. In a blog post titled "Poison Ivy: Breakfast of Champions," Julie Craves noted: "This week, we got two poison-ivy seeds from a Ruby-crowned Kinglet, the first seed samples of any kind we've had from a kinglet. This particular bird was first banded on 7 October, when it weighed 6.3 grams. We recaptured it several more times. On 12 October, it weighed 6.7 grams, on 16 October it was 7.2 grams, and on 18 October it was 7.4 grams. It doesn't sound like much, but that's a 17.5% increase in the bird's original weight."

⁃ ⁃

We owe a good deal to Dr. Judd and the forgotten scientists who spent their careers examining moose entrails and dissecting woodpecker droppings, and also to the dedicated researchers still toiling in poison ivy–laced woodlots today. Because now we're armed with information. *What do animals eat?* Now we have some clues to help us answer that deceptively simple and all-important question. And the crucial information that researchers discover in the droppings of kinglets, wrens, and warblers can enable us to take action to help threatened wildlife species come back—even from the brink of disaster.

CHAPTER 12

Bird Candy

Albany, New York, 1983

"BLUEBIRD!"

The cry echoed through the office, interrupting the drone of the weekly staff meeting. At the magical word, we all dropped our pencils and leaped from our chairs. Calendars and papers went flying as the entire staff bolted from the room and raced to the nearest window. Panting, we gazed out reverently at the bush where a small bird sat quietly preening its brilliant blue feathers.

———⌢———

This is one of my favorite memories from the years I worked at a nature center, staffed by keen ornithologists who (unlike me) really knew their birds. And way back then, bluebirds were a noteworthy event, indeed. A sighting of a bluebird would merit an instant telephone call to Dial-A-Bird, the birding hotline. Birders would converge with their binoculars, hoping for a fleeting glimpse of this rare and lovely species.

But fast-forward this scene some thirty years, to the present day. A cry of "Bluebird!" in a group of birders would produce—well, perhaps not bored yawns, but casual nods and smiles. Today, bluebirds are a fairly common sight.

For centuries bluebirds were one of the most popular birds in America, the traditional symbol of happiness, and a welcome visitor to yard and garden. But in the 1940s, bluebird numbers began to sink, first slowly, then alarmingly. There were many reasons for this decline: subtle factors that one after another began to decimate the friendly little birds. A deadly combination of habitat loss, increasing pesticide use, and competition from other cavity-nesting birds was topped off by several years of severe winters with ice storms. By the 1960s bluebird populations had dropped to frightening levels, and the bluebird was on the express train to the endangered species list. It seemed inevitable that bluebirds would go the way of the passenger pigeon and the dodo. Extinction.

But for once, the worst failed to happen. Bluebird populations are stabilizing, even rising in some areas. Bluebirds have returned from disaster, like little blue phoenixes rising from the ashes.

There are many reasons for this miraculous comeback. Dedicated volunteers across North America put up special nesting boxes for bluebirds and spend countless hours monitoring the nestlings and maintaining the boxes. Pesticide use, while still a huge issue, is at least more targeted than the uncontrolled and widespread spraying of the 1950s and 1960s, when schoolyards, parks, and playgrounds were regularly and indiscriminately doused with pesticides like DDT. And another factor in the bluebird's happy return is poison ivy.

In spring and summer, bluebirds feed on live insects, capturing their prey in open meadows. But what do bluebirds eat in winter?

Insect eaters like bluebirds have a problem when the temperature sinks below freezing. Many birds simply switch hats, so to speak, and become vegetarians, eating seeds and fruits for the duration of the cold weather. Some of our most beautiful and beloved birds do this—robins, cardinals, mockingbirds, and bluebirds. They forgo the dangers and stress of a long migration and stick around northerly locations in spite of cold weather—as long as they can find food.

Contrary to popular opinion, bluebirds and robins don't fly south for the winter—or, if they do, they don't go very far. We always think of the robin pulling a worm out of the lawn as a harbinger of spring, but actually it's the worm that's the spring harbinger, not the bird. As long as there are food supplies available, the birds' thick plumage keeps them warm enough to survive even in subzero temperatures. A recent Christmas count in one small section of an upstate New York county tallied 450 robins and 120 bluebirds.

But of course many other species of birds do migrate south for the winter. Early in fall, migratory birds get ready for the big push south. Some move singly, some congregate in noisy flocks. Some migrating birds voyage mind-boggling distances—thousands of miles. These migrators need fuel—quick-energy sugars and fats—to propel them on their long journey. Instead of high-protein insects, they focus on foods that are high in sugars, lipids (a type of fat), and calories. The juicy berries of several species of native dogwood shrubs are favorites—the sweet berries are sometimes called "bird candy."

Now, you'd think that *any* old bush that had berries on it would be good food for birds. But not all berries are created equal when it comes to nutrition. Many birds feed on the abundant red berries of the nonnative bush honeysuckles. But honeysuckle fruits are the Hostess Twinkies of the plant world. They're filled with low-quality sugars and lacking in essential nutrients, failing to offer birds the high-fat, nutrient-dense food they need to power their long flight. Birds fill up on the sweet stuff, and then miss out on the more nutritious food they should be eating.

Poison ivy berries are definitely not bird candy. They're not high in sugars. And they don't grow in great abundance, but rather in small clusters here and there on vines and spread across the ground. So the big migrating flocks tend to ignore poison ivy. Instead, the plants that offer high-sugar fruits get mobbed in fall—hordes of starlings or blackbirds will sometimes strip the stems of every single dogwood bush. Soon the sweet, juicy fruits are gone: the wild grapes, blackberries, crabapples, and wild cherries are eaten—only a few shriveled leftovers remain. Then the migrants leave for warmer climes. And as the cold deepens, storm clouds roll in. What's left for the stay-at-home bluebird to eat?

You guessed it.

Poison ivy's fruits are often described as "waxy" in texture. They're not juicy and luscious. But they're very cold-hardy—the small, dry berries don't shrivel and rot in freezing temperatures. They're very high in fats—more than 40 percent, although not all birds are able to easily or efficiently digest the thick, waxy fats. Various species of warblers and woodpeckers seem to be especially good at digesting the berries, but dozens of species of birds will turn to poison ivy when the going gets tough. Poison ivy berries may not be as highly nutritious as some other species—but they're a whole lot better than nothing at all.

Candy isn't what you need in a long-term survival situation. Paradoxically, poison ivy, the less nutritious food, is the more important when it's crunch time. In the last ice storm of the season, in the bitter cold of a March blizzard, all the sugary-sweet, high-fat berries are long gone. But poison ivy's bounty is still there, giving life in the starving time, the bitter end of winter.

CHAPTER 13

Preparing for Doomsday

SOME SAY THE WORLD WILL END IN FIRE. SOME SAY IN ICE. AND SOME say the world will end in poison ivy.

Doomsday preppers. That's what they're called. They're the folks who are utterly convinced that the end of the world—or at least the collapse of civilization as we know it—is nigh. And if—when!—Situation X arises, you'd better be prepared, they warn. You'll want to have a good strong fence around your compound—a fence of poison ivy.

Doomsday preppers are ubiquitous on the Internet, and their websites offer help as you chart your personal course for coping with the end of the world. And more than a few preppers suggest using PI as an APD (anti-personnel device). It's more highly recommended by some, in fact, than barbed wire. Because you can always hack your way through barbed wire, but everyone is intimidated by poison ivy.

It's a depressing idea, really, the world ending with all of us taking potshots at each other over hedges of poison ivy. I look up from my computer screen, sigh, and can't help but wonder, why? Why so full of hatred, such mindless animosity toward people?

I don't mean the doomsday preppers. I'm talking about poison ivy.

Why is poison ivy apparently so hell-bent on the destruction, or at least the discomfiture, of the human race? Why does it target humans, when apparently the rest of the animal kingdom can go on their merry way, unaffected by the plant that treats us so cruelly?

Why does poison ivy *want* to make our immune systems go berserk? What did we do to deserve it? Poison ivy has no delicious fruit that we're

Scratching the Itch: Only Us?

Although humans are the only species to have such an extreme dermatitis reaction, it's possible, although uncommon, for other species to be affected, too. Guinea pigs, for example. These rotund little creatures have an amazing biological similarity to humans, which is why they're often used in research as, well, guinea pigs. They're especially similar to us in their hormonal and immunologic responses to allergens and so are often used in testing allergy medications. Scientists in quest of poison ivy remedies have succeeded in sensitizing the stomachs of some unfortunate guinea pigs to urushiol, after removing the animals' protective fur.

Primates, from lemurs to gorillas, are of course related to *Homo sapiens*, and there is anecdotal evidence that some species of apes can display a reaction to poison ivy like their human cousins. In the natural course of things, the range of poison ivy doesn't tend to overlap the range of primates (other than humans), but poison ivy could pop up in unlikely places—zoos, for instance—where a captive animal might encounter it.

And there is much evidence that pet dogs can acquire a poison ivy rash as well, although wild canids such as wolves and foxes apparently do not. Fur, of course, provides a lot of protection, but pet websites have posts by the dozen from dog owners lamenting their pet's itchy stomach or rear end after the dog used a patch of poison ivy as a comfort station. Perhaps the unnatural lifestyle and highly processed food of domestic dogs have broken down some of their natural immunities.

anxious to harvest, no tasty leaves or stems we can make into fiber. Poison ivy has no need to fend us off—in fact, if it weren't so toxic, we wouldn't bother it at all; such a pretty vine could move into our neighborhoods and be welcome. Why so needlessly defensive?

Let's look at the question from a different point of view—that of the poison ivy itself. To consider the universe from the viewpoint of a plant requires a huge shift in our thinking, but doing so brings the green world into sharper focus. Every structure and every chemical produced by a plant has a price tag attached—an energy price tag. To create the complex strings of molecules that make up urushiol, a poison ivy plant has to use up a good deal of the precious energy it has created through photosynthesis. Every bit of energy that goes into creating urushiol is energy that can't be used to grow bigger leaves or longer roots. Why would the plant go to all that extra effort to produce a chemical that so dramatically affects a species that *doesn't* prey on the plant?

Could urushiol have another purpose?

Urushiol is a type of oleoresin, a mixture of an oil and a resin secreted by the plant. Many other species of plants create resins, too, but why? Botanists have long debated the purpose of these sticky substances. Some plant resins have long-lasting and memorable aromas—frankincense and myrrh being two famous examples. Their powerful scent definitely attracts humans and may also have some role in attracting helpful organisms, like pollinators.

But most botanists agree that the main function of resin is defense. Resins are created by plants not to attract, but to repel.

Urushiol is stored in the outer cells of poison ivy leaves, stems, buds, branches, and roots; in fact, it surrounds almost every part of the plant like a sticky, fluid coat of armor. (However, urushiol is not found in pollen, nectar, or the nonliving woody cells inside the older stems.) Whenever poison ivy is nibbled, stepped on, chewed, plucked, or damaged in any way, the urushiol immediately oozes out and starts doing its job—protecting the plant.

But how? Since the rash generally appears hours or even days after contact, it seems to be a singularly inefficient way to deter predation. As you scratch Monday's rash, are you really going to remember exactly which plant you brushed against on that hiking trip last Saturday? But urushiol's job is not to scare away humans. Urushiol is targeting smaller enemies.

—◦—

Over the eons, the chewing mouthparts of insects—uncountable billions of insects—have been the major force that has shaped and guided plant morphology. The evolution of plants is inextricably twined with the evolution of insects, and they've been joined in this duel for some 350 million years. It's a sort of steadily escalating arms race—I'll develop thorns, you develop strong jaws that can munch thorns. Okay, I'll develop *poisonous* thorns. You develop antidotes to the poison. And so on, down the millennia. Botanists theorize that most plant defenses—thorns, poisons, spines, camouflage, scents—have evolved as a way to survive the constant onslaughts of hordes of hungry insects.

Poison ivy is a plant that is heavily impacted by an incredible variety of insect life. Look at any poison ivy vine (not too close) and you'll find dozens of tiny nicks, scars, bites, nibbles, and holes in the leaves. Poison ivy, which claims so many human victims, is itself the victim of relentless predation.

Scientists in a 1989 study were seeking a way to control poison ivy biologically, by discovering an insect that fed on it and could keep the plant in check. The entomologists spent weeks surveying tens of thousands of poison ivy leaves to find out what was eating them, and in the process they discovered that *Toxicodendron radicans* is gnawed on by no fewer than thirty-five different types of beetles, two species of fly, two kinds of wasps, five types of aphids, and thirty-seven species of caterpillars. In addition, more than thirty other species of insects and mites feed on poison ivy.

That's not even counting the dozens of species of insects that use the plant for shelter or egg laying, often damaging it in the process. When an infinitesimal critter called the poison ivy leaf gall mite (*Aculops rhois*) lays its eggs on the leaves, the result is a horribly mottled red surface that looks—poetic justice!—exactly as though the poison ivy plant has just contracted a nasty rash.

Against this onslaught of tiny foes, urushiol is poison ivy's key to survival. When a poison ivy plant is injured, the resin promptly seeps out and fills the wound. The sticky stuff makes it harder for insect jaws to keep munching. It also helps to block the entry of infection—not unlike the way blood works when you cut your finger. But the sticky ooze of urushiol may be even more effective than blood at trapping germs, spores, and bacteria. Many plant resins have strong antibacterial and antifungal properties, and urushiol probably does as well, although scientists have yet to research it for these qualities.

Urushiol is oily and therefore highly waterproof (as many people have discovered from fruitlessly trying to flush it off their skin), so it isn't washed away by rain. The sticky fluid quickly hardens into a tough, resistant layer over the plant's wound, protecting it like a scab on a child's cut finger. It's a remarkably effective botanical first-aid kit—and poison ivy developed this defense long, long before humans came on the scene. By the time of humans' arrival on the planet, poison ivy had already been battling insects for millions of years.

And so that dose of urushiol that bathes our ankles as we tramp through the forest isn't intended for us. It's a mere accident of evolution that urushiol happens to be a powerful allergen for most humans. Humans aren't even on poison ivy's radar, so to speak. We're just collateral damage.

And finally, that's the joke on the doomsday preppers. Poison ivy is all but useless as a deterrent compared to thistles, stinging nettles, or thorny briers, which bring an instant "Ouch!" The only way to use poison ivy as a barrier, I suppose, is to trade on its lethal reputation. Bear this in mind, so you'll be ready when the apocalypse goes down.

Never mind the barbed wire. Just put up a big sign that says BEWARE OF POISON IVY in front of your bunker, and watch 'em run.

CHAPTER 14

Holding On to the Land

Cape Cod, Massachusetts, 1797

THE LITTLE SEASIDE TOWN WAS KNOWN IN ITS EARLIEST DAYS AS DANgerfield. Not because the gentle green fields and pastures surrounding the cluster of cottages were dangerous. Or because the sand dunes fronting the ocean were thickly laced with poison ivy. It wasn't the land, but the wide blue fields of ocean that were deadly indeed. The treacherous sandbars just off the coast of Cape Cod made Dangerfield into the graveyard of hundreds of ships.

The solution, of course, was a lighthouse. In 1797 ten acres of sandy coastland, thick with dune grass, bayberry, and poison ivy, were purchased from one Isaac Small at the then-enormous price of $100. A wooden tower was built, a whale-oil lamp was installed, and on a dark night in 1798, the first beam shot out from the Highland Lighthouse.

Today, technology has come to the aid of ships at sea, and the breakers off Cape Cod are no longer the menace they once were. Dangerfield is known as the pleasant town of Truro, and the Highland Light is merely a tourist attraction. The dramatic golden rays still glow in the night sky—but with one big difference. The lighthouse tower doesn't stand on the same spot it occupied in 1798.

In 1996 the lighthouse was moved—yes, the whole lighthouse was picked up and moved, inch by inch. In an incredible feat of engineering, the tower was pulled five hundred feet back from the water. This multimillion-dollar move was necessary if the lighthouse was to continue to exist, because the land where it had stood for two hundred years was crumbling beneath it, vanishing with every wave that struck the land. Of Isaac Small's ten acres, less than four remained. The rest were gone.

As every high school geology teacher knows, there's nothing more surely guaranteed to make students yawn than a lecture on soil erosion. What could be duller than grains of soil washing or blowing away, bit by bit, over hundreds or thousands of years?

Erosion is a subtle disaster, but it's a disaster nonetheless. The catastrophe of vanished soil has toppled civilizations and plunged nations into chaos. More than two thousand years ago, the philosopher Plato pointed out the disastrous effects of erosion in the ancient world: "What now remains compared with what then existed is like the skeleton of a sick man, all the fat and soft earth having wasted away, and only the bare framework of the land being left." Much later, Franklin D. Roosevelt agreed with him about the importance of soil conservation, remarking that "the nation that destroys its soil destroys itself."

But still we don't get it. We occasionally concern ourselves with plants—corn, or tulips, or poison ivy—but rarely give a thought to the dirt they grow in. At least, until the dirt isn't there anymore. Then a Dust Bowl forms where there used to be farmland, or a desert replaces a rain forest.

How does it happen, this stealing away of the earth? It often begins with a disaster, of which nature has an endless supply. Lightning starts a fire; flooding damages a riverbank; earthquake, tidal wave, hurricane. Somehow the protecting skin of vegetation is torn open. Once soil is exposed, wind and water capture it, speck by speck, down to bedrock.

Barrier beaches like Cape Cod are zones of perpetual disaster. Even on the calmest of summer days, they're relentlessly scoured by waves; during a big storm, beaches are battered and pounded. Something like fifteen thousand waves strike a beach every single day. Bit by bit, the land dissolves. So what glue can possibly hold grains of soil together tightly enough to defy the ocean's hammer blows?

Well, poison ivy.

Very few plants can survive in the barren region of sand, just past the reach of the tide, known as the foredunes. It may be a great place to spread your towel and sunbathe if you're a vacationing human, but if you're a plant, it's more forbidding than the Sahara. No shade. Little organic soil, just a dry, salt-infused layer of sand. High winds. Constant temperature fluctuations—well over a hundred degrees Fahrenheit on a sunny day. Briny spray from the waves. Not the place you'd choose to plant a garden.

The sand dunes of barrier beaches are the most fragile of environments. Unanchored, a heap of sand grains would vanish in the first gust of wind. A sand dune can't form until the sand is held in place by plant roots. As ecologist Janine Benyus says, "Beach plants are the lion-tamers in a circus of wind, water, and waves." Without the wide, branching rootlets of poison ivy and its neighbors, the beach we're sunning on could soon be only a memory.

Beach grasses are usually the very first pioneers to colonize bare sand. Their roots begin the process of holding the shifting sands in place, creating a new dune, or restoring an old one after damage. Shortly after the beach grasses take hold, the next settlers arrive, the hardy, salt-tolerant species: bayberry, beach plum, wild beach roses, sand cherry, and poison ivy.

Plasticity—that amazing ability to adapt to a wide range of conditions—stands poison ivy in good stead here. Poison ivy growing on a beach is a different critter than poison ivy in a rich-soiled, damp, shady forest. In arid environments, poison ivy leaves develop a waxy cuticle over the leaves to minimize water loss. Leaves are smaller to offer less surface area to sun and salt. Roots spread wider and shallower to drink up any passing drop of rain.

Yes, PI loves to lie on the beach, basking in the sun—as many an unwary vacationer has found. The seedlings can't grow in pure sand, but all the poison ivy vine has to do is find a spot where a gull dropping or a crab has decomposed to yield a bit of fertile ground for a seed to sprout.

Poison ivy rapidly goes sprawling over the terrain, soaking up sun to power its rapid growth like a widespread network of green solar panels.

And poison ivy is a fast grower, to put it mildly. A vine can grow six feet in a year. Even more important, poison ivy grows as rapidly below the ground as above. The shallow roots can spread in a circle twenty feet wide. That's approximately three hundred square feet of soil-grasping rootlets for each mature plant. The very qualities that make poison ivy so incredibly hard to get rid of when it's invading your backyard make it priceless when it comes to protecting the land.

The coast guarded by the Highland Lighthouse is today barren beach and crumbling cliffs, as is much of Cape Cod's shoreline. But four hundred years ago, when the Pilgrims disembarked from the *Mayflower*, Cape Cod was a very different landscape. Just as John Smith walked in a Virginia forest that no longer exists, the passengers on the *Mayflower* saw a sight we'll never see—sand dunes that were thickly wooded almost up to the water's edge.

The coastal soil was securely held in place by a vast network of roots—not only beach grass and poison ivy, but a low-lying forest of sturdy, compact trees: scrub oaks, junipers, cedars, and sassafras. A dense understory of beach plum, wild cherry, and bayberry as well as PI made these dunes a rich habitat for wildlife.

However, like most humans, the English settlers were sorely uneducated in the dangers of soil erosion. They and their descendants cut the oaks and cedars, grazed cows on the beach grass, overharvested the beach plums and bayberries, and diligently eradicated the poison ivy. With the result that soon Cape Cod's dunes were heaps of sand crumbling away under homes and lighthouses. Dangerfield's worst danger turned out to be not the sea, after all, but the shifting ground under our feet.

There's an old story (fictional, I fear) of the courageous little Dutch boy who spotted a hole in the dike that held back the ocean. Knowing that the tiny breach would soon widen and lead to a devastating flood, he put his finger in the opening and saved the day.

Nowadays, his heroic role is filled by poison ivy, which is widely planted in Holland to stabilize the sandy, salty soil of the crucial dikes that hold back the sea. The Dutch are especially fond of PI not only because it grows quickly to create a secure covering for the dike, but because its presence tends to discourage trespassers.

Today, the little Dutch boy has his work cut out for him. Centuries of abuse and neglect have made dunes all along America's coast vulnerable, instead of the bulwark against storms that nature intended them to be. Dr. Richard Watson, a geologist and activist for beach conservation, writes that "our natural dune seawall is our only protection from total destruction in a major hurricane." One day, he warns, "we will have our Katrina, or Rita, or Celia, or Carla, or 1919 storm. It is coming, the only question is when. . . . We can ignore powerful natural systems for a while, but in the end man always loses."

Hurricane Katrina, Superstorm Sandy: these powerful ladies taught us what happens when we stop thinking about the soil. Part of the reason that these storms were so devastating was that the beaches they hammered weren't lined with beach grass and cedar, bayberry and poison ivy. They were lined with houses, hot dog stands, clam shacks, hotels, motels, and roads. Development increases the vulnerability of the land in a thousand subtle ways—lowering the ground level, compacting or loosening the soil, obliterating wetlands. And worst of all, removing the anchoring roots that clutch those elusive grains of sand. No motel or restaurant owner wants poison ivy surrounding his or her place of business.

But the fiercer the storm, the more important a natural seawall such as a dune becomes. Ironically, in an extreme disaster, the function of the foredune is to be washed away. Not even the deepest plant roots can guard forever against a hurricane—some erosion is inevitable. During a big storm, sand from the foredunes is washed offshore, which flattens the beach profile. As Dr. Watson explains: "This has the very useful benefit of causing the waves to expend more energy offshore and reduce the rate of attack on the main dune line, buying precious time so that the storm has time to pass before destroying the entire natural dune seawall and the structures behind it." In the fiercer storms that are surely coming our way, poison ivy could help to buy us time.

Nowadays, it seems as if we have succeeded in controlling the natural world. We can predict the weather, build dams and levees, keep ourselves safely insulated from the storm's fury. But as we discovered during Hurricane Katrina, no human-made bulwark can ever be strong enough to hold back the ocean. When it comes to keeping the water out and the land in place, we can strive to build ever higher walls. Or we can let nature do at least some of the work for us. The amazing paradox is that fragile stems of beach grass, cattails, and poison ivy can abate the storm's fury more effectively than sandbags or cement barricades.

From the Bay of Fundy to Cape Cod, down the coast to the Everglades, along the banks of the Rio Grande and the Mississippi, in the coastal bayous of Louisiana, acres of America are anchored firmly in place by poison ivy.

CHAPTER 15

There's Gold in the Hills: Poison Oak

Los Angeles, California, 1769

THE CITY OF LOS ANGELES HAS HAD MANY NAMES IN ITS LONG HISTORY. Sin City, it's sometimes called (vying with Las Vegas for that title). Tinseltown, or the Big Orange. Some sources will tell you that its original name was Nuestra Señora la Reina de Los Ángeles de Porciuncula, given by a Spanish priest in the 1700s. But actually Los Angeles has a far older name than that. Long before it was the City of Angels, the spot was known to the Native Americans who had lived there for millennia as Yang-na, or "place of the poison oak."

Now, "place of the poison oak" might not sound like a spot you would describe as delightful. Poison oak grows abundantly in the area around Los Angeles (the parts that aren't paved). In fact, huge swaths of California and indeed the entire West Coast are covered with the stuff. Poison oak is a highly detested nuisance.

But consider this description of Yang-na, written on a hot August day in 1769 by a Franciscan monk named Juan Crespí, one of the first Europeans to view the area: "We entered a very spacious valley, well grown with cottonwoods and alders, among which ran a beautiful river ... This plain where the river runs is very extensive ... After crossing the river, we entered a large vineyard of wild grapes and an infinity of rose-bushes in full bloom. All the soil is black and loamy and is capable of producing every kind of fruit and grain which may be planted." Crespí wasn't describing a dry wasteland infested with poison oak—on the contrary, he rhapsodized at length about Yang-na, calling it "this delightful place among the trees on the river." But wait—how can a place be delightful if it has a ton of poison oak?

———

What is poison oak, anyway? It's not related to oak trees but has leaflets that are irregularly lobed, rather like oak leaves. It's the botanical first cousin to poison ivy, and there is eastern (also known as Atlantic) poison oak in the southeastern United States, which is considered a different species from western (aka Pacific) poison oak in the western part of North America. Poison oak thrives in sandy soil, and in the east grows especially along the coast from New Jersey to Texas.

Western poison oak flourishes in a wide range of habitats, all along the coastal region from British Columbia in the north to Baja in the south. The stuff is everywhere, spreading over dry hills, climbing eroded canyons, creeping into housing developments and construction sites. And like poison ivy, poison oak is packed with urushiol from root to leaf tip, and can bestow an epic rash on humans who brush against it. Nobody's favorite plant.

But for uncounted centuries, when the Native American tribes of the West Coast considered poison oak, they found a plant of usefulness.

Scratching the Itch: Oak vs. Ivy

Poison oak and poison ivy both have plenty of urushiol, and the rash they bestow is pretty much the same. Botanically, they're closely related, having diverged from the same parent plant, a New World *Toxicodendron*, many eons ago. Their range tends to differ, poison oak being more tolerant of warmer temperatures and nutrient-poor soil. However, the plants' ranges often overlap, and it's very possible to find them growing side by side.

They must have exercised caution when dealing with it, since no one is guaranteed immunity to urushiol on the basis of race (or religion, gender identification, age, or sexual orientation).

But even though the plant has always been a hazard, there is considerable recorded evidence for poison oak being used by West Coast Indians in a host of ways. Interestingly, Native American healers used it much as Dr. Dufresnoy and other European physicians had, brewing a decoction of the plant to treat skin ailments, sores, warts, and so forth.

But there were some other quite remarkable ways that Native Americans made use of poison oak. Anthropologists studying West Coast

Scratching the Itch: How Can You Weave a Poison Oak Basket?

It sounds like masochism to weave a basket of poison oak, no matter how elegant the enamel-black stems look. Using poison oak in cooking seems the height of insanity. And the West Coast Indians certainly weren't all immune to poison oak and ivy. But there is considerable evidence that they built up a resistance to urushiol by ingesting small amounts of poison oak leaves. Eastern tribes may have used the same technique with poison ivy.

Eating poison oak or ivy is obviously fraught with peril, and much has been written about this risky and highly controversial method of preventing urushiol dermatitis (see appendix for a more thorough discussion of its pros and cons). But given the repeated references in ethnobotanical literature to Native American use of both poison ivy and poison oak, perhaps it tended to work.

On the other hand, Thomas E. Anderson, an expert on poison ivy research, notes that perhaps the amazing accounts of spitting salmon on poison oak branches reflect "the credulity of some researchers, and a better sense of humor than they gave their informants credit for."

tribes noted that the broad leaves were used in cookery, to wrap salmon filets while broiling them over the ashes. Poison oak twigs were used for propping open gutted fish while roasting them. Native American basket weavers also used poison oak's sap to color their basketwork. Urushiol, when exposed to air, quickly turns an elegant, shiny black, and the stuff is so waterproof that it makes an excellent dye. Peeled poison oak twigs turn black, again from the urushiol-laden sap darkening when exposed to air, and make a dramatic addition to basketry.

Poison oak, like poison ivy, has immense value for wildlife. Many species of birds feed on poison oak berries. And the plant's oak-shaped leaves are rich in protein, making it a high-quality browse for herbivores. Especially in dry chaparral or desert environments, poison oak is a key wildlife food for deer, rabbits, bear, and elk. Poison oak nectar is nutrition for a wide range of insects, especially honeybees.

So poison oak has always been an important part of the ecology of the American West. The plant flourishes in a wide range of habitats, from the dry canyons of southern California to the dense forests of the Northwest. But, once upon a time, it was just one plant among many others. Poison oak was one small part of a diverse landscape of forests and grasslands, laced with crystal streams and rivers, rich with birds, fish, and all kinds of wildlife. A delightful place, indeed.

Until a balmy January day in 1848, when a carpenter named James Marshall found a nugget of gold at Sutter's Mill.

Within a year, tens of thousands of optimistic miners had flooded into the territory, and another quarter of a million were on the way. The delightful landscape was about to change, big-time. And the gold seekers didn't know it, but they were about to create a paradise for poison oak to grow in.

The early birds, the miners who got there first, found gold pretty easily. All they had to do was swirl a few pans of creek water, or dig into a hillside, and there among the silt and pebbles were grains of gold. But within weeks, the surface gold was tapped out. Now miners had to seek harder and dig deeper.

The first gold had been found in a creek, so waterways were the first places to search. Swirling water in a pan was replaced by the technique of dredging—dragging a bucket along the river bottom. It was a simple, if time-consuming process—just wash through the dredged material to separate out the gold and then dump the silt and mud, called tailings, on the riverbank. Soon lone miners wading in the creek with a handheld bucket were replaced by enormous machine-powered dredges, and piles of tailings seven stories high lined riverbeds.

Then miners began to search for gold on the land, blasting holes and drilling tunnels in mountains. Wagonloads of rock and soil were washed for gold, and vast mounds of tailings dumped. But dredging rivers and tunneling into hillsides was a slow process. A new method, called hydraulic mining, was much more efficient.

Imagine squirting a garden hose on a mound of dirt. The force of the water easily washes the dirt away, right? Now imagine a monstrous hose with a nozzle fifteen feet long, shooting a high-powered jet of water onto a hillside. Hydraulic mining turned out to be a highly efficient way of obtaining gold. It was also an environmental disaster.

One visitor to a hydraulic mining site wrote in 1872: "It is impossible to conceive of anything more desolate . . . than a region which has been subjected to this hydraulic mining treatment . . . Certainly by no other means does man more completely change the face of nature than by this method . . . Hills melt away and disappear under its influence . . . The desolation which remains . . . is remediless and appalling."

Mammoth hoses stripped off the protecting layer of vegetation and blasted the soil away, leaving a wrinkled moonscape of rock. In less than thirty years, hydraulic mining dislodged an estimated eight times as much soil as was removed in the digging of the Panama Canal. Millions of tons of sediment obliterated streams and dammed rivers, leading to devastating floods that swept away trees . . . oh, I'll spare you the rest. It's too depressing. No one wants to read the unhappy story of melted hills, ruined rivers, wrecked grasslands, vanished forests. You get the idea.

The gold rush had turned the "delightful landscape" into a war zone—man battling nature, no holds barred. But eventually, finally, the gold started to run out. (And, I'm happy to say, hydraulic mining was banned in California in 1884.) The frantic pace of mining was slowed, if not halted. And what was left, after the dust cleared? A deeply wounded land, where almost nothing could grow.

Nothing, that is, except poison oak.

The devastation of mining created a congenial habitat for poison oak. It's one of the very few plant species that was capable of finding a toehold on the rock-hard piles of tailings and the eroded remnants of hills. The plant excels at exploiting the most unpromising of habitats, and is well adapted to thrive even in soil that has been depleted of nutrients. Poison oak can grow in places that even poison ivy can't abide.

But in the damaged landscape, poison oak was no longer one plant among many in a diverse and healthy environment. Now a monoculture of toxic leaves blanketed the sadly altered land. Poison oak continues to this day to be the dominant plant covering many thousands of acres of the American West.

Since California's dry landscape has always been extremely prone to wildfires, the bumper crop of poison oak has become a murderous problem. Breathing the urushiol-laced smoke from the burning hills wreaks havoc with lungs. It's a serious hazard to anyone, but to firefighters battling California's increasingly severe epidemic of wildfires, poison oak smoke is a life-threatening risk. Droplets of urushiol can be carried far and wide by the smoke. The urushiol attacks nasal passages and lungs, causing potentially fatal injuries. The effects of inhaled urushiol on the lungs has been compared to the effects of mustard gas on the soldiers of World War I.

Horrid rashes, deadly smoke: nothing to love about poison oak, surely? But poison oak can take an unsuspected role in nursing injured land back to health. Because what do you do with a wound? You bandage it.

Nature's bandages are the plants called seral species. These are hardy, aggressive, fast-growing plants that can cope with the conditions that occur after disaster, even on eroded, depleted soil. Poison oak and poison ivy are both highly effective and widespread seral species.

It's not that seral species *prefer* poor soil, mind you. If you plant poison ivy or oak in rich black compost and then weed out the competing plants and throw in some fertilizer, you'd better stand back. But seral species don't mind sun and can cope with drought. They send out a dense maze of rootlets that pry compacted soil particles apart, while the spreading fingers of rootlets grasp the soil, holding it in place to prevent further injury.

And some seral species, such as poison oak, serve a particularly vital function: they are not only bandages, but nurses.

Much research has been done on the aggressive ways that plants compete with each other. Plants can poison competing plants, shade them out, outgrow or even strangle them. But plants can also benefit each other in surprising ways. Some plants, like poison oak, are considered "nurse plants," facilitating the growth of other species. These helpful plants are of great interest to botanists who are concerned with restoring and revegetating damaged habitats.

Now it's not that poison oak is being nice and trying to help out a poor, struggling little seedling. No altruism is involved here. But a hillside of poison oak can hide some pleasant surprises beneath those much-dreaded leaves.

Nurse plants affect a host of complex and subtle factors: light, temperature, soil humidity, and nutrient content. The nurse plant creates a benign microhabitat, allowing other species to find opportunities to grow in a location where they wouldn't ordinarily be able to survive.

Usually, shade discourages seed germination. But in the desert-like conditions on the eroded hillsides after the miners got through, shade is essential. The broad, oak-shaped leaves of *Toxicodendron diversilobum* act as life-giving parasols against the sun. They shade a tiny patch of soil, creating a haven that is just a bit cooler, moister, more protected from the wind. The poison oak nurse also hides its "nurslings" from the mouths of hungry herbivores.

Then, like any good nurse, poison oak provides nourishment. Poison oak's penetrating roots suck calcium and other minerals from deep in the ground, and carry the nutrients upward to the rest of the plant. As the poison oak leaves die off each autumn, they decompose, breaking down into nutrient-filled compost that enriches the depleted soil. The nurslings now have the sustenance they need. Sometimes, in fact, they grow large and strong enough to oust their nurses. The newly fertilized and loosened soil can be colonized by other plants: grasses are often the next to move in, then shrubs and trees. Slowly begins the creation—or re-creation—of a diverse and healthy ecosystem where once there was an impenetrable thicket of poison oak.

That's how nature heals wounds. The concept of using nurse plants is being investigated all over the world as a way to restore the damage caused by mining, flooding, construction, bomb craters . . . all the ills that nature and humanity visit on the earth. Poison oak is recommended by the US Forest Service (although with appropriate cautions!) for use in land restoration projects.

Poison oak will never, for obvious reasons, be a popular and widely planted species. But the poison oak growing naturally will, if left to do its work, continue to mend the soil and nurse other plants to health. Slowly, slowly the injured land regenerates, under the healing touch of poison oak.

CHAPTER 16

The Devil You Know

THERE'S DANGER HERE.

Severe skin irritation . . . Itching, blisters . . . Flush the eyes or skin with water for fifteen minutes as soon as possible after contact . . . Exposure can cause abdominal cramps, diarrhea, difficulty breathing, headache, nausea, vomiting, low blood pressure, general weakness and possible coma. . . . Oh, dear.

It's an autumn day, bright and crisp, with a still-warm remnant of summer sun, but the ominous words send a shiver down my spine. When I bought the container of herbicide in the garden store, encouraged by a helpful shop clerk, I hadn't bothered to read the fine print. Now, my finger on the trigger of the convenient spray nozzle, I pause and squint to finish reading the instructions—in extremely small print—for calling the poison control center and not inducing vomiting in case of accidental ingestion. Hm.

I glance from the plastic jug of poison in my hand to the poison ivy vine draping the back fence. The three-parted leaves are wearing their Halloween costume of gold and purple, edged with scarlet. Danger lurks there, too, of course—severe skin irritation, itching, blisters—all sorts of unpleasant consequences.

What to do? Okay, I'm willing to cede a goodly portion of my yard to wildlife habitat and native species and preventing soil erosion and all that good stuff—but things are getting out of hand. There's poison ivy twining up the swing set. Poison ivy creeping up the trunk of the apple tree. Poison ivy lurking amid the daffodils and lilacs.

I shift the heavy plastic jug to my other hand, wondering if I should be wearing gloves. Or a face mask. Then the poison ivy leaves rustle in the breeze, as if flaunting their dreadful possibilities, just waiting for my kids to brush up against them.

Life, I guess, is all about choices. Choosing your poison, so to speak.

A backyard or a playground infiltrated by those little three-parted leaves is an obvious invitation to disaster. When humans and poison ivy choose the same plot of land to inhabit, something has to go, and it's not going to be the humans. Agreed. There are places where poison ivy simply must be eradicated, and the question is not *whether* we're going to get rid of the poison ivy but *how*.

According to the advertisements on television and in gardening magazines, the answer to that question is to go to the local garden store and buy a bottle of weed spray. It's cheap, convenient, and easy. Grab, aim, squirt. But on scanning the shelves of herbicides in the garden shop, I found a bewildering variety of possibilities. Asking for assistance, I was guided by a friendly employee toward a handy-size jug of the most commonly used herbicide in the United States—indeed, in the world: a chemical compound known as glyphosate. It's used in schoolyards, playgrounds, campgrounds, backyards, gardens, and most abundantly in crop fields; the US Environmental Protection Agency estimates that thirteen to twenty million acres are treated with 18.7 million pounds of glyphosate annually.

But "glyphosate" is a hard name to remember, harder to spell, and has an unpleasantly chemically sound to it, so the herbicide is usually called something else—a friendlier nickname, so to speak. Just as poison ivy seemed less threatening when Dr. Dufresnoy referred to it as *Rhus*, herbicides containing glyphosate are marketed in the United States, and worldwide, under a variety of muscular aliases: Aquamaster, Bronco, Buccaneer, FallowMaster, Honcho, Prosecutor, Ranger, Rascal, Rattler, Razor Pro, Touchdown IQ. And then there's the trademarked name which is best known: Roundup.

⬿⟋

There's an old saying: Better the devil you know than the devil you don't know. Before I start squirting herbicide—by any name—next to the swings and the seesaws in the backyard, I want to know more precisely what it is that I'm squirting. I'm familiar with poison ivy's potential, for good or ill, but what about glyphosate? Exactly how bad is glyphosate for me if I breathe it, or touch it, or spill it on my clothes? What happens if it percolates through the soil and perchance infiltrates my drinking water? Just tell me the facts, so I can choose the lesser of two evils.

But these are tough questions, and it's hard to get a straight answer. The truth varies, depending on whom you ask. Trying to find out if there are problems with glyphosate is like trying to pin down the wind. Governments, colleges, universities—all of the "authorities" are oddly elusive on the subject. Website after website, including ones maintained by federal and state agencies, abound with vague terms like "toxicity yet to be determined," "not yet fully understood," "has the potential to," "as far as is known," "no more than slightly toxic . . ." Even the EPA seems to be a little confused.

Glyphosate

- Glyphosate products can be safely used by following label directions.
- Glyphosate has low toxicity for humans.
- Protective eye wear is recommended for the few products that may cause eye irritation.
- Entry into agricultural fields is allowed 12-hours after application of these products.
- Glyphosate is no more than slightly toxic to birds and is practically nontoxic to fish, aquatic invertebrates, and honeybees.
- Certain products contain an ingredient that is toxic to some fish.

Source: www.epa.gov

Glyphosate has been in use for decades, and when it first came out its defenders were vociferous. And there did seem to be a lot of evidence to show that the stuff is relatively safe. It breaks down and decomposes in soil (although residues can persist for six months). It seems to have no immediate effects on drinking water. It's not deemed to be terribly toxic to human health, or birds, or fish. (But if it's low toxicity, how come I have to stay out of the field for twelve hours after application? Just asking.)

The EPA has been reviewing the toxicity of glyphosate for the past several years, and as of this writing still lists it as "undergoing registration review."

New research, though, is looking at more subtle long-term effects. Some scientists are beginning to link glyphosate to a scary range of health problems and diseases. An MIT study concluded that glyphosate's "negative impact on the body is insidious and manifests slowly over time as inflammation damages cellular systems throughout the body. Consequences . . . include gastrointestinal disorders, obesity, diabetes, heart disease, depression, autism, infertility, cancer and Alzheimer's disease."

Every chemical that enters our bodies permeates our entire system in a myriad of subtle ways, for good or ill. Glyphosate inhibits the production of various key enzymes in both plants and mammals, including humans. These enzymes play crucial roles in the body, one of which is to help detoxify chemical residues and environmental toxins. Could this pesticide be making us more vulnerable to other pesticides?

And once we start down the herbicidal slippery slope, we discover it's not really as simple as grab, aim, squirt. One squirt or, well, actually several squirts, are effective for killing off poison ivy buds and leaves. But it's more difficult to eradicate thick stems or a big old vine. You have to keep constantly reapplying glyphosate till the vine has used up all the nutrients housed in the roots and stops putting forth new growth. Many garden websites recommend a hefty dose of Roundup once a week throughout the growing season.

Or they recommend reaching for the bigger guns. If Roundup fails, there are more herbicides waiting in the wings. Poison Ivy and Tough Brush Killer, for example, and other brands that use a chemical called triclopyr (short for 3,5,6-trichloro-2-pyridinyloxyacetic acid). It's a brawny and effective killer of woody plant pests: poison ivy, poison oak, brambles, willow shoots, scrub oak, etc. Spray the Tough Brush Killer once (or twice, or three or four times) and the unwanted plant is dead. Gone for good.

But how, exactly, does this stuff in the plastic jug kill plants so efficiently? Just asking.

Tough Brush Killer is a type of herbicide known as a plant growth disruptor. Essentially, the active ingredient, triclopyr, triggers irregular hormone behavior in plants, causing abnormal growth and, eventually, death. The question that I have to ask myself is, what else does it disrupt?

Even if you bend over and paint the herbicide directly on the target plant, it's just not possible to keep trace amounts of the triclopyr from drifting off in the breeze, or washing off in the next shower of rain, or brushing off on the fur of a passing rabbit. Any plants the chemical comes in contact with will be affected, their growth hormones disrupted just like the poison ivy's are. Okay, I'll keep the stuff away from the roses and lilacs, but what else is going to get hit? Triclopyr disrupts the growth hormones of every plant, great or small. It disturbs the normal growth and nutrient cycling properties of mosses, of ferns, of algae. And triclopyr sticks around long after the poison ivy has gone. Residues of triclopyr can persist in soil for weeks, sometimes up to a year.

What else is affected by my squirt of herbicide? A US Forest Service study noted that newly hatched tadpoles of leopard frogs, green frogs, and bullfrogs "died or became immobile" after exposure to triclopyr. What if there's a toad hopping through the poison ivy, or a salamander prospecting for worms in the soil nearby? What does triclopyr do to a cardinal who grabs some poison ivy rootlets to line her nest, or a warbler feeding on insects on a poison ivy leaf?

The cumulative effects of triclopyr and glyphosate and a host of other herbicides "are not yet fully understood." Translation: Who knows? This is the kind of thing that is all but impossible to research conclusively. How can you follow the cardinal off to her nest and monitor the effects of a beakful of triclopyr?

The effects of chemicals can be insidious—we don't even notice the subtle changes they bring about. You spray Tough Brush Killer on the poison ivy near the pond in July, and then next spring you notice that the spring peepers aren't singing as loudly and sweetly as they used to ...

The question, I guess, is: how much collateral damage are we willing to accept? I'll gladly risk disrupting some moss hormones if it will keep my child from suffering a painful dermatitis reaction. I'll do anything, in fact—anything at all—to keep my children safe.

But safe from what, exactly? Which is the greater danger, poison ivy or pesticide?

One reason often cited for eradicating poison ivy from schoolyards or campgrounds, picnic areas or backyards, is the absolute necessity of keeping it far away from kids. But the hazards of many pesticides are still being argued, and one of the hottest areas of debate is the effect they have on children. The National Pesticide Information Center (a cooperative effort between Oregon State University and the US Environmental Protection Agency) points out that "infants and children are more sensitive to the toxic effects of pesticides than adults. An infant's brain, nervous system, and organs are still developing after birth. When exposed, a baby's immature liver and kidneys cannot remove pesticides from the body as well as an adult's liver and kidneys."

There's increasing evidence, for instance, that exposure to pesticides can trigger allergies and asthma reactions in children. It's a diabolical paradox. Could our use of pesticides be increasing our sensitivity to allergens—like urushiol?

The bitter paradox of parenthood. Wouldn't be the first time that parental good intentions led to the very thing we fear. Are we making our kids more sensitive to poison ivy by trying to protect them from it? Well, who knows, right? This phenomenon "is not yet fully understood," and there is "still inadequate evidence to state," etc.

Will getting a smudge of Roundup on my hands kill me? No. Will toddling barefoot across a playground dosed with triclopyr kill my child? No. Will smoking one cigarette kill me? No, of course not. It's a lifetime of exposure to toxins, inexorably, year after year, that has the potential to cause harm.

So it all comes down to a choice, like everything else in life. Do I really want to be spraying a "potent hormonal growth disruptor" around the seesaws and the sandbox?

And—here's another tough question—even if I do make the decision to use pesticides, will they work? Will I succeed in eradicating the poison ivy, poison oak, dandelions, or whatever pest I want to eliminate? Will the risks I take by using herbicides give the desired result—a safe, completely poison ivy–free environment?

Here's a true tale of one town's attempt to eliminate poison ivy.

CHAPTER 17

No Holds Barred

Brookline, Massachusetts, 1933

BROOKLINE IS A SMALL TOWN JUST OUTSIDE OF BOSTON, MASSACHU-
setts—a nice place to live, and a quiet one. An old town, established in
the 1600s, with two brooks for its borders, hence the name. It's shaded
with trees and has parks and gardens; in fact, it's always been a green
place. Andrew Jackson Downing, a famous horticulturist, described it in
1841 as "a kind of landscape garden . . . inexpressibly charming [with]
lanes which lead from one cottage, or villa, to another . . . quite an Arca-
dian air of rural freedom and enjoyment. These lanes are clothed with a
profusion of trees and wild shrubbery."

A pleasant, uneventful place, it would seem—until the summer of
1933, when war was declared.

In that long-ago summer, the people of Brookline noticed with
dismay that there were quite a few clumps of poison ivy in their town.
Several citizens complained to the town authorities, pointing out that
there were in fact a total of fifty-eight poison ivy growths. Fortunately,
various property owners had destroyed forty-one of the growths, and the
citizenry called on the town council to eradicate the rest.

This seemed like a reasonable and easily accommodated request—
after all, there were only seventeen growths remaining, and the town sent
work crews to deal with the problem. They did so to such good effect that
in 1934, there were ninety-two growths.

Clearly, something had to be done. The town got seriously to work, and an intensive effort was made to eliminate the pesky poison ivy, with the result that in 1935, there were 230 growths.

There's a science fiction movie, *Invasion of the Body Snatchers*, in which seed pods of a giant alien plant fall from the sky. They begin to multiply at a frightening rate and eventually take over a quiet little town not unlike Brookline. The film was made in 1956, so this grim vision was still in the future, but still, the citizens of Brookline were beginning to be seriously alarmed. They proceeded to conduct a no-holds-barred, all-out war against poison ivy.

They began by using chemicals, but harmless ones you might find in your kitchen cabinets. Poison ivy plants were doused with solutions of table salt or borax powder. This method killed off some of the foliage but left roots undamaged, and soon poison ivy was energetically resprouting.

The work crews then proceeded to use fuel oil, spraying the growths with kerosene and gasoline, which soaked into the ground. This killed poison ivy, as well as every other nearby plant, but even then, poison ivy returned, sprouting at the edge of the gasoline patches and spreading over the damaged soil as vines do so efficiently.

Next the town turned to the destructive power of fire. Workers were equipped with "flamers" to incinerate stubborn poison ivy growths. This highly dangerous method had been popular for many years, especially out West in battling those huge swaths of poison oak. But it's a dangerous process messing around with flamethrowers. In Washington State an employee battling poison oak was tragically burned to death when a flamer hose burst and sprayed him with gasoline.

Anyway, back in Brookline, the flamers got to work. And sure enough, flamethrowers turned Brookline's poison ivy to ashes—for a while. But just as the Powhatan Indians of Virginia found when they scorched their hunting grounds and farm plots, poison ivy *loves* fire and happily resprouts after a burn. Also, poison ivy seeds have a hard outer coating and can't germinate until they are scarified, or cracked open, a condition that often results from fire. In one study conducted by the US Forest Service, the density of germinating poison ivy seeds was 300 percent higher in burned than in unburned soils. Soon green shoots of poison ivy were rising from the blackened Brookline soil.

Salt, gasoline, and fire having failed, Brookline turned to more complex chemical weapons. The next item in their arsenal was a concentrated solution of copper sulfate. Exposure to this bluish-green compound can lead to a variety of unpleasant symptoms in humans, severe eye damage being one. The chemical is highly toxic to fish and other aquatic life, but it was sprayed on the banks of Brookline's streams and rivers.

But still poison ivy returned. The town authorities decided to get out bigger guns.

It's hard to know how well the effects of sodium arsenite were understood by the unfortunate workers using it. This was in the thirties, remember, and in the pre-Google era, information wasn't so easily obtainable, and people weren't as paranoid about health risks as we are in our more nervous modern times. Most likely, they just sprayed away without giving possible health consequences a thought.

As its name suggests, sodium arsenite contains arsenic. Ironically, for a substance used to battle poison ivy, even low doses of sodium arsenite cause itching, skin irritation, and rash. Prolonged exposure to sodium arsenite causes nausea, vomiting, diarrhea, and convulsions and may lead to nervous system damage resulting in eventual paralysis, and death. It's also considered a dangerous carcinogen, and a teratogen, or cause of birth defects.

By 1942 poison ivy was seemingly as hard to eradicate as the Nazis, but the citizens of Brookline were as determined as the Allies on D-Day. The next weapon in the arsenal was ammonium sulfamate, a substance toxic to birds. It's another chemical that can cause skin irritation in humans, as well as a variety of other symptoms. Brookline's defenders used ammonium sulfamate in such high concentrations that it corroded the metal spraying tanks unless they were scrubbed and rinsed after each use. Each poison ivy growth area was sprayed three times, at forty-eight-hour intervals.

In 1945, in that triumphant and victorious spring, Brookline claimed victory. Poison ivy had been eliminated within the town borders.

The *Brookline Tab* is the community weekly newspaper of Brookline, Massachusetts. In 2013 a guest columnist wrote an article entitled "The Plague of Poison Ivy" for the Garden section.

A newcomer to Brookline, he had recently moved into a new house, and he described his ongoing battle to eradicate the poison ivy that surrounded it. "It started when I began clearing the land to our new home and farm in early summer. Volunteer trees and poison ivy vines, so dense that in most areas you could only see a few feet into the thicket, overtook the property. These beasts certainly took full advantage of their several years of solitude and unrestrained growth, after the previous owners downsized and before the house was finally occupied again, by me. . . .

"For the past six months and well into the future, my focus has been and will be mostly on the eradication campaign of these godforsaken vines. But it is now bordering on obsession. I'm seeing progress and even find myself looking for small windows to get outside during the week to go at it some more."

He may not realize how hard it is to fight history. Poison ivy was just one plant among many in a healthy ecosystem, before humans declared war on it. Once the flamethrowers and the lethal chemicals had destroyed the protective groundcover of native plants, the door was thrown open for poison ivy to move in and become the "beast" that took over the neighborhood.

CHAPTER 18

The Future of Poison Ivy

ONE THING THE DOOMSDAY PROPHETS ARE RIGHT ABOUT: THINGS ARE changing, for sure. There's a lot more poison ivy in our future.

There is actually much more poison ivy on the planet today than there was in the time of John Smith and Pocahontas. In the seventeenth century, poison ivy had to wait for a fire or a flood to lay the earth bare, so that it could outcompete taller, more established plants. The number of places where poison ivy and poison oak could elbow their way in were fewer.

Nowadays, it's a lot easier to find disturbed soil. Anywhere the vegetation is peeled away, poison ivy can jump in. Construction sites, for example, where they're building that new condo or strip mall. Long corridors under power lines where the power company thoughtfully removes the tall vegetation that would compete with poison ivy. Or along the eroded edges of hiking trails—ironically, there's often much more poison ivy along the edges of trails, ankle-high, than there is in the surrounding forest where the soil hasn't been disturbed.

Yes, poison ivy loves us—or at least loves the kind of environments humans create: frequently mowed roadsides, lawn edges, and playgrounds. Overused picnic areas and campsites. Telephone poles. Forests fragmented by logging roads, channelized riverbanks, wilderness subdivided into housing lots. Your backyard. Some botanists estimate that the amount of poison ivy on the planet has doubled in the last fifty years.

Poison ivy takes advantage of all these opportunities not because it's an evil plant, but because it's a species with tolerance for a wide range of conditions. Outcompeting less hardy and adaptable species, it exploits every niche available to it and soon grows overabundantly. It sounds, in fact, a lot like us humans.

━ ⁓

Several years ago, researchers at Duke University in North Carolina were curious to see what would happen to a forest under increased levels of carbon dioxide—the kind of air we may well be breathing in the not-too-distant future. They used a system of carbon dioxide–pumping pipes to create high atmospheric CO_2 levels in a series of forest plots: two hundred parts per million higher than the current norm—a level that many scientists predict could be a reality by 2050. Then for six years, the researchers patiently watched while poison ivy grew. And grew and grew.

Every part of poison ivy—fruits, stems, and roots—became more abundant. And the leaves were larger, too. The roots were more vigorous and wide-spreading. Poison ivy growth increased, in fact, by approximately 150 percent.

In a similar experiment, in which scientists increased the amount of CO_2 available to poison ivy, researchers (presumably wearing gloves) slowly removed leaves from poison ivy plants to simulate the effects of animals eating them. They found that the leaves grew back far more quickly with the extra dose of CO_2.

And not only was there more poison ivy, it was *better* poison ivy—better from the plant's point of view, that is. Because the urushiol in the CO_2-enhanced plants was increased, both in volume and in toxicity. Poison ivy on steroids, so to speak.

In a report entitled "Rising Atmospheric Carbon Dioxide and Potential Impacts on the Growth and Toxicity of Poison Ivy," L. H. Ziska and coauthors noted that "Increasing CO_2 resulted in significant increases in leaf area, leaf and stem weight, and rhizome length and biomass. The amount of urushiol produced per plant increased significantly for all CO_2 above the . . . baseline. Significant increases in the rate of leaf development following leaf removal were also observed with increasing CO_2."

━ ⁓

So, yes. Poison ivy is increasing and spreading. But not because humans are encouraging it—far from it. We spend billions of dollars annually on poison ivy control and cures. Yet poison ivy ignores us and goes on its merry way.

And there's very little doubt that poison ivy, and its cousin poison oak, will continue to increase. An excess of poison ivy is part of the bitter price we pay for disregarding nature's rules.

So perhaps the doomsday preppers are right. Bad things are coming down the pike. Better get ready. Poison ivy is getting stronger, bigger, and badder. Maybe we need to try harder to get even—to get rid of it at all costs. Surely, with all the power of modern science at our fingertips, we can even find a way to eradicate poison ivy for good.

But here's the question: should we?

Let me digress a bit and mention another pesky plant, often targeted by Tough Brush Killer, a species of tree called honey locust. It's a native American species, like poison ivy, but it's unpopular in backyards, parks, and gardens because it has thorns. I mean big thorns. Not just little prickers—great, huge whomping thorns, up to six inches long and iron hard. Old-time carpenters actually used them as nails. Quite a nasty plant.

But nowadays, no worries: there's a thornless variety of honey locust, especially developed by horticulturists for areas where humans might come into contact with it. Honey locust defanged, as it were. This thornless honey locust is the classic tree of big-box shopping malls, planted throughout urban/suburban areas because it's hardy and fast growing. It seems practical and convenient to lose the thorns. Don't want sharp thorns in an area where kids might prick themselves.

But here's the interesting part. Honey locust, being a native species, pops up not infrequently in Native American stories and legends, probably because the wicked spines made the tree a useful bad guy. Indian tales, like all folklore, often have what anthropologists call a "test theme"—a situation in which the hero or heroine has to be tested somehow, prove something, achieve something. In a long-forgotten Cherokee myth about Thunder Spirit, the dreadful spines of honey locust are the test.

Ethnographer James Moody recorded this version of it: "The news came to Thunder that a boy was looking for him who claimed to be his son. Said Thunder, 'I have traveled in many lands and have many children. Bring him here and we shall soon know.' So they brought in the boy, and Thunder showed him a seat and told him to sit down. Under the blanket on the seat were long, sharp thorns of the honey locust, with the points all sticking up, but when the boy sat down they did not hurt him, and then Thunder knew that it was his son."

But the remarkable thing is that the dangerously sharp spines were essential in order to perform the crucial test—to weed out the special from the ordinary, to recognize a powerful spirit. The thorns were, in a way, sacred. But Thunder would be out of luck today, when all the honey locusts have branches smooth as a baby's bottom.

It's an arresting thought. What do we lose when we take the thorns off the honey locust? Or when we put up guardrails on a steep, high mountain ledge? The result is safer, yes—but it's not quite the same view. The world is less threatening and scary, for sure, if we get rid of all the prickers—if we eliminate the wolves and cougars, spray the wasp nests, round up the poison ivy.

This is an especially important concept for those who argue that we must protect children from poison ivy at all costs. If poison ivy is something that never enters children's experience, if it's something that's not remotely a part of their worldview, then a devastating surprise may await them on their first venture outside the controlled confines of the playground. For so many people, an unfortunate encounter with poison ivy forever slams the door on any possibility of loving the outdoors.

———

Usually a class of fifth graders bounce and chatter like a flock of starlings on a spring day. But this group shuffled behind me, talking among themselves in low murmurs. The closer we got to the head of the trail, the quieter they got.

A girl with purple leggings and a pink top sidled up next to me. "So are we, like, going into the woods?" she inquired. Her nametag had "Vanessa" written on it with swirls and curlicues around the letter *V*.

"Yep," I said.

Vanessa looked doubtfully at the green line of trees ahead. "Like, way in the woods?" she asked.

"Way in," I agreed.

"Oh," she said, biting a fingernail. "I've never been in the woods."

"Never?"

"Well, we went in a little way once, but my mom got poison ivy."

Not long ago, I was an educator at a small suburban nature center. Every spring, big yellow buses would converge on the place, bearing school classes for their annual field trip. My job was to lead the groups of students on nature walks. And over the years I worked there, I noticed something. Every year, it got harder and harder to convince kids to go into the forest.

Oh, they were still enthralled by the birds and trees and the rippling stream, once they got started—but increasingly, it was hard to convince them to leave the parking lot behind, no matter how many times I assured them everything was perfectly fine. Of course, assuring a kid that everything is "perfectly fine" doesn't work. Wary from years of vaccinations, dentist appointments, and standardized tests, kids know just how trustworthy that statement is. But adults never learn.

"Oh, it'll be perfectly fine!" I told Vanessa breezily. "Don't worry, I'll show you just what poison ivy looks like. It has three leaves so it's easy to identify."

"My God, you don't have poison ivy here, do you?" The high-heeled chaperone, just behind us, turned pale. "Oh, my goodness, boys and girls, I had poison ivy once, and I'll tell you, it was terrible. My skin got all red and itchy and the blisters were as big as eggs, and . . ."

"Yes, well, here we are at the trailhead," I broke in. "We'll start the nature walk now."

"Now remember, boys and girls, stay in line," the chaperone hissed. "Keep your hands at your sides and don't touch anything. Once you get poison ivy, it spreads all over, and there's no stopping it. Now keep a sharp eye out for plants with three leaves."

"Three leaves, right." Vanessa frowned at a maple tree. "But there's three leaves on this branch, and three leaves on that branch—and look, that branch over there has five or six leaves. So this must be poison ivy."

"Just don't touch *any* leaves!" the chaperone insisted. "And stay right in the middle of the trail. Only the ranger knows what's safe to touch. I tell you, the blisters were the size of golf balls . . ."

No wonder kids like to stay inside and play video games. At least indoors you know what's safe to touch.

Children can only learn what they're allowed to experience. And in case you haven't spent any time in a classroom recently, I'm sorry to report that children aren't being taught the ABCs of the forest. There are no standardized tests on watching woodpeckers or cardinals, no pop quizzes on how to find salamanders. It seems to me that Vanessa—and all those thousands of children standing fearfully in the parking lot—need some help in learning how to go out in the woods and mess around and climb trees and watch tadpoles and all that stuff that textbooks leave out.

And here's what really scares me. The other day, I took a bunch of kids on a nature walk, and when I mentioned that they should keep a watchful eye out for poison ivy, they didn't say, "Poison ivy! Oh, no!" One and all, they innocently asked "Poison ivy? What's that?" Countless children don't even know that such a thing as poison ivy exists. Walking in the woods and fields is not something they ever plan to do, you see, so they don't need to know the hazards—just like I don't need to know the hazards of taking a stroll on the planet Mars. I'm never going to go there anyway, so why bother?

A more important goal than eradicating poison ivy, it seems to me, and perhaps a harder one to achieve, is learning to coexist with it. And a huge part of that task is teaching our children to recognize it. It's a parent's greatest challenge: making them aware of the dangers of the big bad world, and then crossing our fingers as they go out there to explore. Ignorance breeds fear. It's the knowledge of what's out there, lurking in the tall grass, that can give us the courage to stray off the sidewalk.

EPILOGUE
To Feed a Mockingbird

Albany, New York, 2015

THEY'RE CALLED MOCKINGBIRDS BECAUSE THEY IMITATE THE SONGS OF other birds. But it's more than just imitation—the mockingbird takes the basic notes of other birds' calls and mixes them up, combining tweets and warbles and quavers into a unique harmony. It's like listening to a great jazz musician improvising—creating a melody that's never been heard before, never to be repeated. This remarkable music is the reason for the famous old saying that it's a sin to kill a mockingbird.

In the twenty-first century, I'm happy to say that it's not only a sin but a crime to kill a mockingbird. They're protected by state and federal law, and harming them is punishable by a fairly hefty fine. But if it's a sin to kill a mockingbird, it's surely an even worse transgression to starve one to death.

———

One afternoon, I pull into a shopping mall parking lot. It's crowded because of Presidents' Day sales, so after a certain amount of fruitless cruising, I end up in a parking spot way in the back. There's a small strip of—well, I can't call it a forest—a tiny scrap of unmowed ground with a few weedy trees and a tangle of bushes and vines.

It's certainly not a national park or anything. Usually I wouldn't even glance at it. But I'd rather do almost anything than shop, so I sit behind the steering wheel, idly looking it over. Even in the dreary end-of-winter mud, I recognize the crumpled stalks of milkweed, blackberry brambles, a few gangly pokeweeds. There's a little wild cherry tree, its naked branches draped with vines. It's utterly unremarkable, the usual mess of unnoticed weeds that fringes schoolyards and malls, highways and housing developments across the country.

I'm sitting there, gearing up for the mall, when a bird flits over the sea of cars and darts into the tangle. I'm no great birder, but I'm pretty sure it's a mockingbird—a big gray bird, with long graceful tail and a white flash under the wings. *What's it doing here?* I wonder, glancing around.

Not so long ago—I remember it—this was a meadow, a mockingbird-friendly sea of waving grasses and berry bushes. Now it's a massive spread of blacktop, with an ocean of cars parked nose to tail. In the chill of late winter, there are precious few opportunities in this barren landscape for a mockingbird—or any creature—to find nourishment for soul or body.

Except for this weedy little strip, where the mockingbird is perched in the spindly tree. Then it hops to a vine—a hairy-looking vine that winds its way up the side of the tree like a furry snake. The odd, fuzzy vine has a few stalks with little white berries on them. The mockingbird starts gobbling them down.

———

"In wildness is the preservation of the world," Henry David Thoreau famously remarked. He's often misquoted as having claimed that *wilderness* was the essential ingredient, but no—*wildness* is what he said, and what he meant. Great naturalist though he was, Thoreau spent surprisingly little time in any place you could call wilderness; as he said, he had "travelled a good deal in Concord," passing most of his life exploring the weedy bits of nature in his hometown—the woodlots and hedgerows and mini-forests tucked between the houses and along the railroad embankments. Even at Walden Pond, he was close enough to civilization that he could pop home for dinner any time he felt like it. He wanted to live "a primitive and frontier life"—a *wild* life—"though in the midst of an outward civilization." Thoreau knew that wildness, like the kingdom of God, is found within you.

Why do we need wildness, after all? There's the aesthetic need, the spiritual craving for peace and greenery and crystal waters. But there's a colder, harder, more practical sense in which Thoreau's quote is true. Humans can pull off astounding technological miracles, we can go to the moon or restructure DNA, but the one thing we can't create is wildness.

Hungry birds? Why not just put up a birdfeeder? But birds, and all wildlife, need more than sunflower seeds. They need a wide range of wild, native plants for food and shelter in order to survive. Trouble is, hundreds of species of native plants are shrinking in numbers. Habitat disruption has changed the chemistry and composition of the soil in many areas. Problems like habitat loss, overpopulation of species from deer to earthworms, invasive plants, and imported fungi and diseases have fundamentally altered the American landscape. Key wildlife food plants like oaks, hickories, wild raisin, wild cherry, blueberries, pokeweed, walnuts, beeches, and a host of others are no longer found in the numbers they used to be. Poison ivy—tough, hardy, and widespread—is rising in importance as a stopgap emergency food for stressed animals, especially in winter.

The legendary American wilderness that John Smith and Pocahontas, Bartram and Audubon knew is all but vanished. We'll never see the forests of beech trees mobbed by passenger pigeons, the cypresses eighty feet high, the dunes thickly protected with fragrant sassafras and cedar. For most of us, what remains of wilderness is generally experienced on the glossy pages of a calendar, or on a few brief vacations. The nature we live with, day by day, comes in bits and pieces, backyards and roadsides, town parks and small forgotten chunks. Like this little island in the parking lot.

I sit and watch for a few more minutes. The bird ignores me and continues chowing down on the poison ivy—it's a good-size vine, with quite a few berries still left on the twigs. Then there's a whisk of gray tail as a squirrel darts up the cherry tree. A crew of small dust-colored birds are hopping around under the poison ivy, kicking up dead leaves as they seek out hibernating insects. Under the tangle of vines and shrubs are hiding places for mice, meadow voles, or maybe a chipmunk; perhaps there's enough shelter for toads or tree frogs. Dried milkweed stalks show that this will be a promising, if tiny, habitat for monarch butterflies come summer. This scrap of wildness is brimming with life.

And it occurs to me—what if we cherished such places?

It's so easy to walk right past this little speck of wildness. But what if it wasn't ignored, but lovingly tended? Just as we might conscientiously stoop to pick up a piece of litter, what if we all learned to yank out a stalk of garlic mustard springing up along the edge of the parking lot. Or carry a penknife to cut bittersweet that's twining around an aspiring young oak tree? Or left a swath of our backyards unmowed, a haven for clover and asters? We can't create wilderness—we can't build a Yosemite or a Grand Canyon in our neighborhoods—but what if we nurtured wildness wherever it sprang up? What if this little speck of wildness were turned into a mini bird sanctuary? A monarch butterfly rest stop, a chipmunk haven. Imagine thousands of these little islands, scattered in parking lots and backyards all across the United States, fitting into each other like pieces of a giant jigsaw puzzle.

＊＊＊

As I get out of the car, the mockingbird flaps its wings and lets out a harsh "chip!" with a glare in my direction. Mockingbirds defend their food sources fiercely, and this mockingbird is protecting his food and his future—the poison ivy vine. Mockingbirds prefer to nest not in deep forest, but in shrubby areas, meadow edges, and even in little thickets like this one. Perhaps when spring rolls around he'll find a mate, and they'll build a nest under the shade of the toxic three-parted leaves. There's plenty of room to raise a family in the comforting embrace of poison ivy.

So maybe the doomsday preppers are wrong. Maybe fear and hate and poison ivy won't be the end of the world, after all.

Maybe it all begins with poison ivy.

Acknowledgments

I owe thanks to many people who weren't afraid to wade into the poison ivy with me.

My sincere gratitude to my agent, Regina Ryan, whose effective editing and creative ideas were of great help in shaping the book.

I am extremely grateful to Daniel Capuano, Fred Breglia, and Alan Mapes for reading the manuscript and giving me the benefit of their expertise. Especially, many thanks to Stephen Young for his thoughtful and helpful suggestions.

I am much obliged to Linda Marshall's encouragement, and to Bob Marshall, MD, for medical insights.

Thanks to Gabriela Lamy of the Chateau de Versailles for unearthing an amazing number of references to poison ivy's regal history.

Thanks to Julie Craves of the Rouge River Bird Observatory and the University of Michigan for permission to quote from her fascinating research and for use of photographs.

I am grateful to Nancy Castaldo and Lois Miner Huey for much helpful advice and good cheer.

Any errors in the book are mine alone.

As always, my sincere thanks to my family. May they always enjoy the beauties of poison ivy while escaping its perils.

Appendix

How to Avoid, Heal, Obliterate, and Coexist with Poison Ivy

PART 1: HOW NOT TO GET POISON IVY

I CAREFULLY PICK THE TINY RED LEAF. As I HOLD IT BETWEEN FINGER and thumb, the leaf catches the early spring sunlight like a ruby. I hesitate, take a deep breath. Should I try it? My first bite of poison ivy.

It's research, after all. There's a centuries-old tradition that eating a leaf of poison ivy every day in the spring will make you immune to the itch. It must be true—I read it on the Internet. Various websites attribute the custom to French Canadian loggers, Spanish conquistadors, or "the Indians."

But it's not just Internet nonsense. Several more reputable sources agree that it's possible to attain immunity by eating poison ivy. Euell Gibbons, the famous wild foods guru of the sixties, swore by this method—he recommended starting in spring when the leaves are small, and eating one a day, therefore getting a small and slowly increasing dose of urushiol for your system to develop an immunity to. I once met a young man who had tried this system—he assured me that not only did it work to perfection, but the poison ivy leaves were delicious, with a "light, lemony taste."

The early spring leaf in my fingers is no bigger than a pea. How could such a small amount hurt anyone? But research has shown that a sensitive individual can get a horrendous reaction from fifty micrograms of urushiol—and, just for comparison, a grain of salt weighs about sixty micrograms.

I drop the leaf on the ground. Nope. Just can't do it. I'm not particularly susceptible to poison ivy, but allergies are tricky things. You can develop them as you age, or grow out of them. A nibble of poison ivy might heighten my immunity, or it might serve to sensitize me so that my next poison ivy encounter would be worse than ever. I guess I'm just not a truly dedicated researcher.

Short of dining on poison ivy, are there any other ways to combat it? The modern medical profession can supply us with steroids and other drugs to combat the itch once we've acquired it, but not all of modern medical science has so far come up with a method for infallibly preventing it.

Before You Touch Poison Ivy

In order to combat something—in order to thwart an enemy, and possibly turn it into an ally—we need to know how it works. Just how does this little green leaf work its devastation on the human race?

Many people swear that they've gotten poison ivy simply by walking past it. Not so. This is a long-cherished belief, and researchers in the nineteenth century spent years trying to isolate the deadly "effluvia" that floated through the air. But no part of poison ivy is windborne—not even the pollen, which is physically picked up and carried from place to place by pollinators, mostly bees. You can't get a rash just by being in the vicinity of poison ivy. In order to develop the dermatitis reaction from poison ivy, you have to come in direct contact with the bad stuff: urushiol.

What Is Urushiol?

Urushiol is a complex combination of chemicals that is actually found in dozens of species of plants. It's abundantly secreted by the plant in tubes called resin canals, which are located in the phloem, the layer of living tissue just under the bark. Poison ivy's sap carries urushiol throughout the plant, from leaf tip to root tip. Urushiol is found in the fruit and flowers, but not in the actual wood cells of the vine.

Resins tend to be sticky substances. If you've ever parked your car under a pine tree and tried in vain to scrub the pine resin off the hood, you'll know what I mean. Urushiol is just as tough to remove from your skin, as it's an oleoresin, both oily and sticky.

Urushiol is, actually, beautiful. Inside the plant it's transparent, but as soon as the chemical hits the air, it combines with oxygen and darkens. Another name for poison ivy is markweed, because the sap made an excellent ink. It was often used for labeling linen, as the dark letters wouldn't wash off, even in the boiling water of old-time laundry vats. (A handy tip for when you're heading off to summer camp—label your underwear with the sap of poison ivy.) I would suspect that this system of laundry marking caused an enormous number of undiagnosed rashes in unlikely places.

Urushiol is so beautiful, and so durable, in fact, that in Asia it's been used as a very expensive lacquer for centuries. A plant called the lacquer tree (*Toxicodendron vernicifluum*), native to Asia, is tapped for its sap, like a sugar maple. Sadly, lacquer tappers enjoy their work considerably less than do maple-sugar makers. But the lacquer made from the urushiol is a lovely, gleaming ebony black; it is waterproof and lasts for centuries.

But for most of human history, urushiol was a puzzle. No one knew exactly what about those weird *Toxicodendron* plants caused dermatitis. It was a Japanese chemist, Dr. Rikou Majima, who finally identified the culprit. In the early twentieth century he spent years studying the sap of the lacquer tree and established the toxin's chemical makeup. He named it urushiol from *urushi*, the Japanese word for lacquer.

This courageous researcher (I hope he was immune) discovered that urushiol is mixture of organic chemical compounds known as alkyl catechols. These long, stringy molecules start with a ringed structure of hydrogen and oxygen followed by a long tail of carbon atoms called the alkyl chain. In different species of plants, the urushiol differs slightly in molecular structure.

Urushiol molecules penetrate the epidermis (the thin outer layer of skin) and then combine with proteins found in the dermis, or inner layer. The urushiol/protein combination is an *antigen*, or a substance that provokes an immune system response. White blood cells recognize these urushiol-infested proteins as invaders and ring the alarm bell, so to speak, activating the body's immune defenses.

The thing with poison ivy is—it isn't the first time that gets you. The first time you encounter urushiol, your immune system responds promptly—but not overtly. You're unaware of the complex internal processes going on inside your body as your immune system identifies the chemical. Lymphocytes (white blood cells), once sensitized, are programmed to respond to future urushiol attacks, and these "memory cells" can persist in the body for many, many years.

As always with poison ivy, everyone's reaction is a little different. Some people can encounter poison ivy multiple times before finally the immune system decides, *okay, time to get serious,* and your body develops the dermatitis reaction. Some people get a horrendous rash on the second exposure. Others don't react at all. You never know.

For those of us who are looking to get outside and enjoy nature—casual strollers or serious peak baggers—are there measures we can use to guard ourselves? Short of locking ourselves in the house all summer, what are we to do? How can we wander ankle-deep in the beauties of nature and not come home with poison ivy?

Well, there are obvious precautions, such as wearing shoes and long pants instead of flip-flops and shorts. If you crave bare skin (and I admit that during a heat wave it's hard to endure long sleeves and pants), there are several excellent products on the market that act like sunblock—creams that form a physical barrier between the toxin and your skin.

But the best way to avoid poison ivy—the only surefire, foolproof way—is to know what it looks like. Unless you're a firefighter or a telephone lineman on an emergency call, poison ivy is completely avoidable. It won't sneak up behind you and bite you on the leg. You just have to keep an eye peeled for it. That's all—nothing easier, right? Just be aware of what plants surround your every step on earth. But I sometimes think that's the real reason poison ivy exists—to jolt me into mindfulness when I walk through the woods without awareness.

It is, to be sure, a difficult plant to identify. Like snowflakes, no two poison ivy leaves are exactly alike. Sometimes the leaves are the size of a dinner plate, sometimes smaller than a dime. The plasticity that serves poison ivy so well, enabling it to adapt to an incredible range of habitats and conditions, means that a poison ivy plant in Texas may look very different from a poison ivy plant in Connecticut. Color varies, too: shiny maroon in early spring, pinkish-green in late spring, inconspicuous green in summer, and a rainbow of golds and purples in autumn.

I've often heard it said that poison ivy is shiny. As ever, it varies; the leaves can be dull green, often marred with insect bumps and bites. But especially in early summer, I've seen poison ivy shining like silver in the sun. A waxy coating helps keep moisture in the leaves and can reflect sunlight.

How to Identify Poison Ivy

Here are some maxims, probably centuries old, that sum up the most important identification clues:

Leaflets three, let it be!
Like clover, three leaflets make up one leaf. Poison ivy leaves grow alternately up the stalk, unlike the leaves of many other plants, which grow opposite each other.

Hairy vine, a danger sign!
Aerial rootlets along the side of the vine help poison ivy climb trees and give the vine a furry look.

Berries white, take flight!
Small, waxy fruits are often hidden under the leaves, green at first, then whitish. Not all poison ivy plants have them.

After You've Touched Poison Ivy

What if you've gone and done it—brushed up against a vine, waded through a cluster of poison ivy? Time is of the essence. The best hope to avoid a rash is to wash off the oil as soon as you can.

Use lots and lots of water—cool, not hot. Hot water opens the pores and allows the urushiol to sink deeper into the epidermis. If you're far from civilization, even splashing around in a stream or taking a dip in the lake will help. Flooding the area with water right away is the best chance to clear off the urushiol. So swim, shower, hose down, do whatever you can to get the oil off your skin.

Should you use soap? Experts differ. Soap can help shift the oily urushiol but also strips your skin of its own protective oils. If you're done hiking for the day, a brisk scrub with soap is probably a good idea. Authorities also differ on using alcohol to remove urushiol. It can help remove the oil, but like soap, alcohol also strips the skin of its natural oils, leaving it vulnerable if any speck of urushiol touches it.

Gasoline, kerosene, or lighter fluid are also sometimes recommended for washing off urushiol. Don't. Do. It. While it's true that these fluids could help remove urushiol, the hazards far outweigh the benefits. Toxins are quickly absorbed through the skin, and these products are highly toxic. Gasoline rubbed directly on skin can *cause* redness, rashes, and itching—it's a classic case of the cure being worse than the disease. Not to mention the risk of setting yourself on fire.

It may be too late anyway, no matter what preventive measures you take. The urushiol forms a chemical bond with the cells of your skin, and after a while (variable—anywhere from a few minutes to an hour or so) the urushiol is so tightly bonded that it is impossible to dislodge.

So, once you've washed off as much urushiol as possible, then what? There's nothing to do but wait and see what develops . . .

Part 2: How to Treat Poison Ivy

Okay. It's quite possible that you've picked up this book, checked the table of contents, skipped all the other stuff, and turned straight to this page. You've got an itchy rash. You don't give a rodent's posterior about poison ivy's botanical nomenclature and horticultural usage. You don't care if Marie Antoinette adored it and Thomas Jefferson planted it in his garden. You just want the itch to STOP RIGHT NOW.

Let's get down to business. To soothe the itching, get wet.

Some people swear by a cool bath or shower, others by a hot one. Cool or tepid water soothes the itch for a few moments. For longer-lasting relief, though, try a hot shower, hot as you can stand it. Seems like hot water is the last thing you would want to soothe a burning itch, but heat does several things.

First of all, heat causes tiny blood vessels in your skin to widen, bringing more blood to the area. This can cause the red rash to get temporarily redder and slightly more swollen. But the increased blood flow is a powerful aid to healing. As Lisa Shea, author of a helpful website on skin issues, puts it, "It's like having an accident on a highway and having a traffic jam—then suddenly doubling the width of the highway so more ambulances can get in."

Even more important for short-term itch relief, heat overrides the constant message of *itch itch itch* that is flowing along your nerves to your brain. A new message—*hot*—diverts the brain, directing its attention to a potential hazard. It takes a while for the itching to start up again. Some people claim a really hot shower can relieve the itching for several hours.

Whatever the temperature, you can't go wrong by taking a shower or bath. There is no urushiol contained in the blisters, so you won't spread the rash by showering, bathing, scrubbing, sweating, scratching, etc.

Obviously, be careful not to use water so hot that you burn your skin. I've found that many PI websites and experts ban hot showers, but they're warning against taking a hot bath *before* the rash develops—which is indeed a bad idea. Hot water opens your pores wide and lets them drink in even more urushiol. That's why you want to wash off with cool water, not hot, immediately after contacting the plant. *After* the rash develops, hot water can be sooooo nice.

The Rash

Some people find redness appearing on their skin within minutes of touching poison ivy. This is unusual and would indicate a highly sensitive individual. It's usually longer before you realize the awful truth. Sometimes the symptoms show up in as little as half an hour, but in other cases, the appearance of the rash can take days. As usual, poison ivy is charmingly inconsistent. The time of onset is determined by how long it takes your immune system to respond, which varies considerably with the individual.

Tiny blood vessels in the skin swell as your system begins to battle the intruder. This produces a sore, itchy rash. Scratching only makes this worse because it causes more irritation and inflammation.

Sometimes it's more than just a rash. Frequently, large liquid-filled blisters may pop up, often in a line showing where the plant dragged across the skin. The blisters may burst and ooze, which is unsightly and makes friends and relatives flee. But (I'll say it again) no urushiol is contained in the blister fluids, so scratching doesn't spread the irritation, as many people claim. A poison ivy rash, however unsightly, is not contagious.

It's not uncommon for people to develop a later, additional rash in a different spot, but it isn't from the blisters, or from poison seeping through your bloodstream or perspiration. It's that the urushiol is still lurking, not inside your body but someplace around you. Perhaps on the shoes you wore the day you walked through that innocent patch of greenery. Or on the dog. Or maybe you washed a pair of jeans together with your underwear, and the urushiol (highly resistant to soap and water) got on your brassiere or jockstrap . . .

The urushiol in poison ivy is an incredibly stable and persistent substance. A trace of it on a shovel handle or a boot lace can give an unsuspecting gardener a rash year after year.

If you find yourself sporting a rash, here's what you *don't* want to do: Google cures for poison ivy. At least not without a healthy dose of skepticism. Remedies offered by the Internet are as numerous as they are mind-boggling. Please, do yourself a favor and avoid any website that mentions lighter fluid. Dried pig excrement is also unlikely to have a beneficial effect. There's almost nothing that hasn't been suggested as a cure for the intense itchiness caused by poison ivy. Mud. Urine. Vinegar. Vodka. Teardrops. Scotch. Joint compound. WD-40 oil. Bourbon. Witchcraft. Prayer.

The most effective cures all have one thing in common—they gently dry out the skin without irritating it additionally. Never cover the rash with any sort of bandages—that will only keep the damaged skin moist, possibly leading to infection. While it seems as though a thick, soothing salve would help, you want to dry the skin, not moisturize it. Oils are not helpful, so avoid lotions or creamy soaps. Gently apply a substance that has drying properties—calamine lotion and witch hazel being a couple of classic cures.

Let air and sunlight do their healing work, and wear as little clothing as you can decently manage over the affected areas. Or you could get into . . .

The Jewelweed Controversy

Perhaps no poison ivy cure has as much controversy surrounding it, so many passionate defenders and equally passionate debunkers, as jewelweed, one of the many traditional Native American remedies for soothing the rash. Jewelweed is a forest wildflower with orange-gold blossoms that dangle underneath the leaves like earrings. Hummingbirds love the cup-shaped flowers rich with nectar. Jewelweed is a native plant, by the way, and I've been trying for years to lure it into my garden so that I can lure more hummingbirds.

But the really remarkable thing about jewelweed is that it completely cures poison ivy. Or not.

The thick stems of jewelweed are soft and juicy. Just grab a handful of jewelweed and rub it wherever you have poison ivy, and the rash will definitely go away. Maybe. In fact, just spreading jewelweed sap on your skin will prevent poison ivy from ever bothering you. Guaranteed. Absolutely. Unless it doesn't.

Jewelweed is Dr. Jekyll and Mr. Hyde. Some swear by its magical medicinal powers and in fact charge a hefty fee for jewelweed products to cure PI. Others scoff. Many experts claim jewelweed has no benefit at all.

Not just on the Internet. Even in antique dusty volumes of herbal medicine from past centuries, jewelweed is hotly debated as a cure for poison ivy. Its detractors say that any benefit comes purely from the fact that its thick stems have a clear watery juice, which may be of some help in washing off the urushiol if applied promptly, but that water would work just as well.

The reason there are so many opinions is that chemically speaking, poison ivy is not a poison with a specific antidote. Urushiol is an allergen, and each person reacts to it differently, and therefore remedies are responded to differently. Allergies are notoriously unpredictable and the effectiveness of remedies can vary immensely.

So jewelweed probably helps some people and is completely ineffective for others. Nothing works like magic on poison ivy, not even the modern medical arsenal of steroids and antibiotics. Poison ivy is too unpredictable, too complicated a question for any one answer to be the solution.

If you're feeling adventurous and want to try it for yourself, how should you use jewelweed? The easiest way is to grab a handful and rub it on. It's that simple. The stems have a clear sap that's very like water and feels refreshing and soothing on the skin. You can chop up the stems and pour hot water over them to make an infusion, and the amber liquid will be a soothing skin wash that, if it doesn't work miracles, at least won't do any harm. You can even freeze the liquid in an ice cube tray, and then you have a refreshing cool rub for sore skin. At best, perhaps jewelweed's mystical properties will work on you. At worst, you'll be no worse off.

A List of Remedies

Besides tramping through the forest in search of jewelweed, are there any other possibilities? Here, in no particular order, is a list of oft-tried remedies for the itch, blisters, soreness, and general nastiness of a poison ivy rash.

- **Calamine lotion.** The well-remembered pink stuff from Grandma's medicine cabinet. A tried-and-true remedy that's been around since the mid-nineteenth century. A blend of zinc oxide and iron oxide, the cool pink liquid gently dries the skin and lessens itching. The classic remedy for mild cases.

- **Baking soda (sodium bicarbonate).** Similar in action to calamine lotion, it's a useful and gentle soother. Make a paste using a little water and dab it on the affected area. You'll leave a trail of fine white powder behind you everywhere you go, however.

- **Swimming in the ocean.** This sounds like one of the pleasanter ways to soothe a rash. Salt water gently dries out the skin. Avoid oily sunscreens, though, and don't sunbathe on the beach—sunburn on top of poison ivy is too awful to contemplate.

- **Scratching.** Don't do it. Don't do it. Don't do it. Scratching makes the skin sore and more sensitive. Clip your fingernails in case you scratch absentmindedly or in your sleep.

The Healing Power of Plants

A tropical jungle, a New England vegetable patch, a citrus orchard, or a suburban lawn—even the unlikeliest environments can provide the means of calming a poison ivy rash. Many plants have strong antibacterial and antimicrobial properties to fight infection, as well as being rich in nutrients like vitamin E and antioxidants to promote healing. This is why Native American traditional remedies, Appalachian folklore, and old wives' tales are often surprisingly effective—they're based on a strong bedrock of science.

Try slices of cucumber or lemon, laid tenderly on your itchiest parts. The inside of a banana peel rubbed gently on the skin is a cure that many people swear by.

A poultice is an old-fashioned medical aid—basically, it's a wad of warm, moist leaves. Pour boiling water over a few handfuls of leaves, let steep for a few minutes, and place the leaves (when cool enough) on the rash for half an hour or so, letting the skin air-dry afterwards. Poultice plants used for generations by both American Indian and European healers include dandelion, plantain, and sweet fern. Personally, I've had very good luck with common plantain (*Plantago major*).

Recovery

Healing time varies with the individual, depending on so many factors: age, nutrition, the strength of the immune system. The rash doesn't actually heal over; instead, the damaged epidermal layers dry up and the skin flakes off, a speck at a time. After a week or so white blood cells start to rebuild the damaged tissue below the rash, and healthy new skin replaces the old.

For most of us who suffer from poison ivy, the best cure is a tincture of time. Or you could try a traditional folk remedy for poison ivy: mix the juice of one plug of chewing tobacco and the juice of three lemons, rub the mixture on the skin, and wait twenty-one days. Since the blisters usually fade by themselves in three weeks, this is one remedy that definitely has a chance of working.

Contact a doctor if:

- The rash covers a large or sensitive area, especially face, neck, or genitals.
- The smoke from burning poison ivy has been inhaled.
- Excessive scratching is making the skin sore, especially if there is any pus suggesting infection.
- Things don't improve after a week to ten days.
- It's urgent to contact the doctor when there is swelling of the throat or any trouble breathing, weakness or dizziness, or severe redness or swelling.

PART 3: HOW TO GET RID OF POISON IVY

I once saw some wise advice, on a bumper sticker of all places—two short, simple words: *Try Softer.* This neatly sums up an idea of immense help in coping with poison ivy, the concept known as IPM: integrated pest management. IPM means using the least harmful alternative to solve a pest infestation problem, whether the unwanted target is dandelions, mosquitoes, poison ivy, whatever. *Try softer.* After all, when you see a fly, you don't get out the AK-47. You reach for the flyswatter.

So what's the "flyswatter" for an invasion of poison ivy? Here's a sampling of some more-or-less effective alternatives we can try instead of immediately rounding it up with herbicide.

Dig it up. Armor yourself well and prepare to dig the roots to a depth of ten inches. Remember that poison ivy's rootlets can spread to a diameter of twenty feet. In general, digging up poison ivy is an exercise in frustration, although perhaps it's worth a try, if there's one lone plant and you feel the need of some healthy exercise. But digging or pulling or any physical attempt at removal is tough and will likely end with many roots remaining in the ground (soon to resprout) and you licking your wounds.

Cut it out. Clip vines at the base, and let the poison ivy twining up the tree die a natural death. (But remember, even the dead plant parts can still cause a rash for a long time.) The problem with this method, of course, is that it leaves the roots in the ground. They will vigorously resprout and sneakily cover the ground with an ankle-high carpet of leaves. Might be better to leave it on the tree—at least you know where it is.

No mow! Where poison ivy is a ground cover, don't, for heaven's sake, mow it, or even worse, weed-whack. There's nothing more disastrous than chopping up poison ivy leaves into little urushiol-dripping bits and flinging them into the air.

Cover it up. Like any green plant, poison ivy will eventually perish if denied sunlight. However, determined poison ivy sprouts will work their way through even a foot or more of wood chips or other organic mulch. Black plastic is probably the most effective way to cover up unwanted plants—spread sheets of plastic over the ground and then exercise patience. Leave the plastic down for a full growing season, at least. A year is better.

Get a goat. Poison ivy is excellent pasturage for livestock: cows, sheep, horses, and especially goats. Indeed, grazing animals often seem to seek it out, passing up tender clover and grass in favor of a juicy cud of poison ivy. The closely related poison oak was found to have 35 percent protein in new-growth foliage, which well exceeds the protein content in most grasses, so you can feel good about encouraging animals to eat the stuff.

The problem, of course, is that having a free-range goat might be unpopular with the neighbors, as well as terrible for the landscape. Goats famously eat everything, tulips and endangered wild orchids as well as poison ivy, so caution is called for here. But for some situations, renting a goat may be the solution. Yes, indeed. There are commercial businesses that will rent an enthusiastic crew of goats, who will eradicate dense stands of poison ivy or other invasive plants that couldn't otherwise be got rid of without a huge dose of pesticide. The goats wade in, and in a matter of hours the PI is gone. (Again, vigilance is required to keep it from resprouting.)

It's an idea that has been around for a while—the US Forest Service has used goats to clear firebreaks since the 1930s. It's a concept that is really catching on, too—considerable research is showing that goats can really make a difference.

Weed control is a particular problem in or near wetlands, because so many pesticides are toxic to fish, amphibians, or invertebrates like dragonflies. A 2014 Duke University study showed that rotational goat grazing (I know, it sounds hilarious) reduced invasive plant cover from 100 percent to 20 percent. The researchers' findings support "an emerging paradigm shift in conservation from high-cost eradication to economically sustainable control." In other words, goats are not only less toxic than glyphosate, but also cheap. They can be rented from local farmers, the kind of small local business it's so great to encourage.

Surprisingly, for a plant that has such an impact on humans, relatively little research has been done on biological ways to control poison ivy. Recently, researchers at Virginia Tech's College of Agriculture and Life Sciences have found a fungus that destroys the plants' seeds. It may be possible to develop a nonchemical herbicidal method of control for poison ivy using the fungus *Colletotrichum fioriniae*.

And all of these methods of poison ivy destruction are fine with me. Much as I appreciate poison ivy's many benefits, I don't object to goats munching it. I cheerfully kill poison ivy myself: I'll happily smother it with plastic, or decapitate it with my hedge clippers. Nor do I frown on any other organic means of clearing out monocultures of poison ivy. Because too much of anything is bad.

A monoculture of poison ivy is just as unhealthy for the environment (or almost) as an infestation of Tatarian honeysuckle or Asian bittersweet. An uninterrupted tangle of poison ivy smothering trees, shrubs, and grasses is a symptom of a serious problem—an imbalance in nature that cries out to be corrected.

Diversity creates a healthy environment. A sprinkling of poison ivy through a woodlot, along a riverbank, or over a sand dune is natural, but it should be blended with all sorts of other species to provide a wide range of food and shelter opportunities for native wildlife. Poison ivy is like salt—a little on the meal is dandy, but a plate full of salt is lethal.

In the end, trying to get rid of every leaflet of poison ivy is impossible. Whatever choice you make, to use herbicide or to go organic, any victory will be a temporary one. Although you might win a battle or two (probably at heavy cost to yourself), you're never going to win the war.

It's like determining you're going to eradicate every single dandelion from your lawn. Banish squirrels forever from your birdfeeder. Get your teenagers to clean their rooms. It's an exasperating struggle that can lead to a kind of tearing-your-hair-out madness. Just when you think you've finally, finally gotten rid of poison ivy in the hedge, or along the trail, or in the corner of the yard . . . all it takes is a single blue jay to lift its tail and excrete a single poison ivy seed, and you're right back where you started.

We need, of course, to set reasonable boundaries—places where poison ivy simply won't be tolerated. And then, as with the squirrels and teenagers, learn when to sigh and shrug, and let nature take its course.

PART 4: TOXIC RELATIVES

It would be bad enough if poison ivy were the only species that contained that maddening urushiol. But the same toxic sap runs relentlessly through the veins of a whole family of plants. Like any family, poison ivy's relatives are a wild array of characters—some inconspicuous, some showy, some harmless, and some pretty nasty indeed, all bound together by shared characteristics and a similar evolutionary history. Poison ivy itself is part of the New World branch of this large and complex family tree.

And as with most families, there are a few surprises. The juicy mango—the sweetest, most fragrant, most luscious fruit on earth—is one of poison ivy's close relations. The cashew, possibly the most delicious nut ever, is in the poison ivy clan, as is the delightful pistachio. Pleasure always has a price, it seems. Mango peels, for example, contain urushiol—not the exact same chemical that's in poison ivy, but very similar. It's enough to trigger a rash in someone who is already sensitized to poison ivy. Fortunately mangoes aren't the sort of thing you bite into, unpeeled, but still they can cause a nasty case of swollen lips. The shells of cashews also contain urushiol, which is why (I always wondered) cashew nuts are never sold in the shells.

Poison ivy's motley crew of relations inhabit all sorts of diverse environments; they're found in Brazilian rain forests and Indian jungles, on Chinese mountainsides, and in Japanese gardens. It's a big clan, to be sure: an enormous family of plants that have been known since the 1700s as the *Anacardiaceae*, which is also called the sumac family or the cashew family. The family has thousands of members, mostly tropical plants, and fortunately not all of them contain urushiol. The *Anacardiaceae* (now there's a spelling bee word if ever I saw one) are divided into many groups (or genera), among which is poison ivy's genus: *Toxicodendron*.

But of course scientific nomenclature of both plants and animals is as changeable as skirt lengths or hairstyles, and *Toxicodendron* species are especially debated. Botanists over the centuries keep on endlessly tweaking the classification, dividing and subdividing genera, changing their minds as to which plant should be in which genus.

Our American Cousin: Poison Sumac

Poison ivy's nearest botanical relatives are poison oak and the deadly little tree known as poison sumac.

Poison sumac is the odd man out, the reclusive cousin who lives deep in the woods and never shows up at the family reunions. It tends to lurk in swamps and wetlands where humans are less likely to be hanging out, preferring riverbanks, ditches, bogs, anywhere it can have wet feet, so to speak. It's a denizen of the Appalachians, the barrens of New Jersey, the Louisiana bayous.

It has some interesting folk names: poison tree, Virginia dye tree, and thunderwood. In all my outdoor explorations, I've bumped into poison sumac only a couple of times. They were memorable times, since poison sumac, at least for me, is poison ivy raised to an exponential power. The urushiol it contains is slightly different than that in PI, and for most people it's much more irritating.

Poison sumac looks nothing at all like poison ivy or oak, so even if you're vigilant about the whole "leaflets three, let it be" thing, it's an easy plant to brush up against by mistake. It's not a vine or a ground cover or a shrub; it's actually big enough to be considered a tree and can grow to thirty feet tall. Instead of the three leaflets, it has multiple leaflets on a stalk, making it look a bit like an ash or elderberry tree.

Confusingly, there are several other plants named sumac: staghorn sumac and smooth sumac are two common ones. These have red berries, which are fabulous winter food for birds, so I'm always sad when these beautiful little trees are eradicated from park or lawn on suspicion of being poisonous. The leaves look a bit like poison sumac, but remember Linnaeus, and take a peek at the reproductive structures: poison sumac has whitish-yellow berries, like poison ivy. Harmless sumacs have red berries. *Berries white, take flight; berries red, no need to dread.*

Just Your Basic Poison Ivy: *Toxicodendron radicans*
In the twentieth century, botanists have split the *Toxicodendrons* into smaller and smaller groups. Now many botanists consider that *T. radicans* should be divided into two subspecies. There's *Toxicodendron radicans (L.) Kuntze ssp. radicans*. It's found in Canada, Mexico, and almost everywhere in between, in a wide variety of habitats.

And then there's *Toxicodendron radicans (L.) Kuntze ssp. negundo (Greene) Gillis*. The L. refers to Linnaeus, the original namer. (One reason for the frequent renamings might be that the renamer still has the honor of attaching his own name to that of the plant.) Subspecies *negundo* occurs mostly west of the Appalachian Mountains. The two subspecies can hybridize, causing botanists infinite headaches as well as unpleasant rashes.

The Other Poison Ivy: *Toxicodendron rydbergii*
And yes, there's another species of poison ivy. *Toxicodendron rydbergii* is referred to in some sources as western or northern poison ivy. Early in the twentieth century, it was separated from its cousin *Toxicodendron radicans* and declared a separate species, named after Per Axel Rydberg, a Swedish-born American botanist. It has a more northerly range than *T. radicans*, flourishing on mountaintops and well up into Canada. The two

plants are all but identical, except that Rydberg's lacks the aerial rootlets that enable *T. radicans* to climb trees. *T. rydbergii* spreads as a ground cover and sometimes grows tall and thick enough to be considered a shrub. Its urushiol content and effects on humans are indistinguishable from those of *T. radicans*.

Endnotes

Introduction

For information on the potency of poison ivy and its effects on humans, I consulted the research of William Epstein, MD, professor of dermatology at the University of California, San Francisco. Dr. Epstein is an internationally recognized expert on poison ivy.

The American Academy of Dermatology estimates that 85 percent of people are allergic to urushiol. Its website (www.aad.org) has much useful information on coping with the dermatological effects of poison ivy.

Chapter 1: The Poysoned Weed

Captain John Smith's works can be found in the Virtual Jamestown Archive, a collaborative digital archive supported by Virginia Tech, the University of Virginia, and the Virginia Center for Digital History at the University of Virginia. See www.virtual-jamestown.org/johnsmith.html. The complete works are also available in print from editor Philip Barbour.

I especially referred to Smith's *A True Relation by Captain John Smith*, published in 1608; *A Map of Virginia with a Description of the Countrey, the Commodities, People, Government and Religion*, published in 1612; and his *Generall Historie of Virginia*, published in 1624.

I found David Price's book *Love and Hate in Jamestown* to be an invaluable source of insights into the character of John Smith. Smith wrote extensively about Native American customs and showed a willingness to learn from them (he learned many words of their language, for example), and I think it's not unlikely that he was open to learning about their use of medicinal plants as well.

For more on Smith's explorations into the Virginia landscape, see the bibliography entries for Susan Schmidt, Lyon Gardiner Taylor, and George Percy. Percy was a fellow colonist who described the landscape altered by Indian fires.

For more on what America looked like before the coming of the Europeans, see Charles Mann's *1491*. He discusses how the well-organized confederation of tribes in the Jamestown area practiced agriculture and regularly burned areas to improve the land for berries and to open up new crop fields.

More information on the effects of fire on poison ivy is found on the website of the US Forest Service. The USFS maintains an online database of plants with much useful ecological information, including their value in controlling erosion and restoring soil, how the plants are affected by fire, and their interrelationships with wildlife.

See www.fs.fed.us/database/feis/plants/shrub/toxspp/all.html#BotanicalAnd EcologicalCharacteristics.

My description of the vanished woods of Virginia is based partly on Donald Culross Peattie's *A Natural History of Trees of Eastern and Central North America*. Writing in the 1940s, he was able to glimpse bits of virgin forest. His descriptions of redbuds, silverbells, and towering sycamores bring to life a little of the vanished woodland that Smith saw.

The name Pocahontas has been translated in almost as many ways as there are historians. "Laughing and joyous one" is from a very interesting oral history of Pocahontas passed down for many generations. See Custalow Linwood and Angela Daniel's *The True Story of Pocahontas*.

For Native American remedies for poison ivy, see the bibliography entries for James Herrick and Gladys Tantaquidgeon. Also see the Native American Ethnobotany website maintained by the University of Michigan: http://herb.umd.umich.edu.

To learn about the Algonquin language I consulted the website of the Algonquin Way Cultural Centre, which lists *mitashkishin* as meaning "infected with poison ivy." See www.thealgonquinway.ca/English/word-e.php?word=910. Poison ivy itself is known as *mitashkishinowin*.

Quotes

1 "Virginia is . . . are unknown." Smith, *A Map of Virginia*.
1 "Much in . . . English Ivie." Smith, *Generall Historie of Virginia*.
2 "Being touched . . . lastly blisters." Smith, *Generall Historie of Virginia*.
2 "Great guilded hopes." Smith, *Generall Historie of Virginia*.
4 "After a while . . . further harme." Smith, *Generall Historie of Virginia*.
4 "Heaven and earth . . . delightsome land." Smith, *Generall Historie of Virginia*.
4 "Cypress trees . . . an Apricock." Smith, *A Map of Virginia*.
6 "A plaine . . . made it." Smith, *A Map of Virginia*.
6 "Were it fully . . . industrious people." Smith, *A Map of Virginia*.
6 "Great smokes . . . down the grass." Percy, "Discourse of the Plantation."
8 "Were assalted . . . Pistoll shot." Smith, *A True Relation*.
8 "Up the River . . . us kindely." Smith, *A True Relation*.
9 "I call them . . . my best content." John Smith, quoted in Schmidt, *Landfall along the Chesapeake*.
9 "Concerning the entrailes . . . for certainty." Smith, *A Map of Virginia*.
9 "Many excellent . . . living Creatures." Smith, *A Map of Virginia*.
9 "Yet because . . . ill nature." Smith, *Generall Historie of Virginia*.
10 "Feare and trembling . . . selfe thereout." John Rolfe, letter to Sir Thomas Dale, 1614, quoted on Encyclopedia Virginia website: www.encyclopediavirginia.org/Letter_from_John_Rolfe_to_Sir_Thomas_Dale_1614.

CHAPTER 2: A COLLECTION OF RARITIES

The website of the Garden Museum in London, UK, has much information on John Tradescant, including the manuscript of *Musaeum Tradescantium*, a catalog of the objects and plants in Tradescant's Ark. See www.gardenmuseum.org.uk/page/our -collections.

The text of *Musaeum Tradescantium* is also online at https://archive.org/details/ musaeumtradescan00trad.

A thorough account of the life of John Tradescant (and his son, John, also a naturalist) can be found in Jennifer Potter's *Strange Blooms* and Prudence Leith-Ross's *The John Tradescants*.

In *The Flowering of the Landscape Garden*, Mark Laird gives a listing of plants in Tradescant's 1634 inventory.

Many sources agree that approximately 15 percent of humans are immune to poison ivy. See the website of the American Academy of Dermatology: www.aad.org.

Quotes

14 "Anything that is strange." John Tradescant, letter to secretary of the navy, in Potter, *Strange Blooms*.

14 "Marchants from . . . known to us." Ibid.

15 "The filth . . . thick ink." Platter and Busino, *Journals of Two Travellers*.

17 "All kinds . . . a plumstone." Georg Christoph Stirn, quoted on the website of the Ashmolean Museum in London: www.ashmolean.org/ash/amulets/tradescant/ tradescant03.html.

17 "A piece . . . dies of it." Ibid.

18 "Great treasurer . . . rarities." Johnson, *Cottage Gardener*.

CHAPTER 3: BARTRAM'S BOXES

John Bartram's garden is now a forty-five-acre National Historic landmark operated by the John Bartram Association in cooperation with Philadelphia Parks and Recreation. See www.bartramsgarden.org.

For biographical information on John Bartram, see the bibliography entries for Edmund Berkeley Jr. and Dorothy Smith Berkeley.

For insights into Bartram's personality, see Thomas Slaughter's *The Natures of John and William Bartram*.

Linnaeus was enthralled by Bartram's discoveries and enthused that Bartram, whom he never met, was the "greatest natural botanist in the world." This was high praise from the conceited Linnaeus, who usually reserved that title for himself. Linnaeus bestowed the name *Bartramia* on a genus of moss.

Andrea Wulf's *The Brother Gardeners* recounts the difficulties encountered by botanists such as John Bartram trying to ship plants across the stormy North Atlantic.

According to Jane Boyd and Joseph Rucker, John Bartram's son William continued his father's work of shipping seeds and plants to Europe. William listed more than two hundred "American Trees, Shrubs, & herbs," including "12 oak species, 9 different pines, and 3 kinds of plums, along with such flowering plants as sunflowers and morning glories. Numbers 114 and 120 on the list, bracketing 5 varieties of grapes, were a bit different: '*Rhus vernix*' and '*Rhus radicans*,' known to us today as poison sumac and poison ivy."

Quotes

22 "I have seen . . . he pass'd by." Richard Stafford, 1668, quoted in Pavord, *Naming of Names*.

23 "Do they think . . . Carolina." John Bartram, quoted in Berkeley Jr. and Berkeley, *Life and Travels of John Bartram*.

23 "The Botanick . . . untill death." John Bartram, quoted on Bartram's Garden website: www.bartramsgarden.org.

25 "Whatsoever . . . beautiful to mee." John Bartram, quoted in Hoffmann and Van Horne, *America's Curious Botanist*.

CHAPTER 4: LOATHESOME HARLOTRY

Solanum caule . . . racemis simplicibus translates as "the *solanum* with the smooth stem which is herbaceous and has incised pinnate leaves." Information on the complex history of classification is found on the website of the Natural History Museum, London: www.nhm.ac.uk. The museum's online Clifford Herbarium database also has images of many of Linnaeus's original herbarium sheets from Clifford's garden, including poison ivy specimens. See www.nhm.ac.uk/research-curation/scientific-resources/collections/botanical-collections/clifford-herbarium/database/browse.dsml.

There are many biographies of Linnaeus. The one I found most helpful is Wilfrid Blunt's *The Compleat Naturalist: A Life of Linnaeus*. Several other biographies, including those by Florence Caddy and Dietrich Stoever, were written in earlier times and are more hagiography than biography, but they contain a wealth of excerpts from Linnaeus's letters.

See Andrea Wulf's fascinating book, *The Brother Gardeners*, for information on Linnaeus and the debate over the classification of poison ivy.

Clifford's garden is now part of the National Herbarium in the Netherlands and is still being used for scientific research. See www.george-clifford.nl/UK/hc_UK.htm.

The Swedish Museum of Natural History has an online display of many of Linnaeus's type specimens, including poison ivy. See http://linnaeus.nrm.se/botany/fbo/r/bilder/rhus/rhustox1.jpg.

The Biota of North America Program (BONAP) has a wealth of information on plants across the continent. BONAP is the result of decades of work by Dr. John Kartesz, along with many collaborators and contributors, including numerous federal and state agencies, arboreta, and universities. BONAP's goal is "to provide all the most current taxonomy, nomenclature, and biogeographic data for all vascular plants and vertebrate animals (native, naturalized, and adventive) of North America, north of Mexico." It includes distribution maps and is a tremendously valuable resource for native plant advocates. See the website for botanical information and range maps for *Toxicodendron* species: www.bonap.net/NAPA/TaxonMaps/Genus/County/Toxicodendron.

Quotes

27 "Barbarian jargon!" Carolus Linnaeus, letter to Albrecht von Haller, June 8, 1737, quoted in Caddy, *Through the Fields*.

28 "Bewitched . . . tulip trees." Carolus Linnaeus, quoted in Blunt, *Compleat Naturalist*.

28 "Truth ought . . . by observation." Linnaeus, *Philosophia Botanica*.

29 "Husbands . . . same bed." Linnaeus, *Species Plantarum*.

30 "What man will . . . without scandal?" Johann Siegesbeck, quoted in Stoever, *Life of Sir Charles Linnaeus*; and Jönsson, "Reception of Linnaeus's Works."

33 "Husbands and wives . . . different houses." Linnaeus, *Species Plantarum*.

35 "If you do . . . lost, too." Linnaeus, *Philosophia Botanica*.

35 "We know less . . . marked for preservation." Wilson, *Biological Diversity*.

35 "Loathesome harlotry." Johann Siegesbeck, quoted in Blunt, *Compleat Naturalist*.

CHAPTER 5: THE BIGGEST BOOK

The engravings from *Birds of America* are viewable online through the Digital Research Library of the University of Pittsburgh: http://digital.library.pitt.edu/a/audubon.

For more on green blindness, see William Allen's "Plant Blindness." He cites Wandersee and Schussler's definition of plant blindness as including an "inability to appreciate the aesthetic and unique biological features" of plants and "the misguided, anthropocentric ranking of plants as inferior to animals, leading to the erroneous conclusion that they are unworthy of human consideration."

Quotes

38 "The inability . . . human affairs." Allen, "Plant Blindness."

38 "Plant blindness . . . condition." Ibid.

38 "As I grew . . . at evening." John James Audubon, quoted in Ford, *Audubon by Himself*.

39 "I know . . . our birds." Ibid.

39 "The nature . . . its inhabitants." Ibid.

40 "To represent . . . and moving!" Ibid.

42 "Impudent . . . stupid book." George Ord, quoted in Souder, *Under a Wild Sky*.

43 "Erect, and . . . of teeth." John James Audubon, quoted in Graham Jr., "Audubon's Legacy."

43 "How could I make . . . large books?" John James Audubon, quoted in Ford, *Audubon by Himself*.

CHAPTER 6: STRONG MEDICINE

Quotes

47 "Spread all over . . . scratching." Joseph Dufresnoy, quoted in "Dufresnoy on Rhus," *British Journal of Homoeopathy*.

47 "I took some . . . considerable doses." Ibid.

49 "Afford a . . . in medicine." Bigelow, *American Medical Botany*.

50 "The *Rhus* has . . . an hour." Scudder, *Specific Medication*.

50 "Valuable medicinal . . . roasted crushed root." Tantaquidgeon, *Study of Delaware Indian Medicine*.

51 "An act of humanity to a royalist." Thomas, *Universal Dictionary*.

52 "How are . . . see them!" Joseph Dufresnoy, quoted in Boyd and Rucker, "No Ill Nature."

CHAPTER 7: ROYAL COLOR

For information on the lives of royal poison ivy fans, see the bibliography entries for Carrolly Erikson, Frank McLynn, Stefan Zweig, and Julia Blackburn.

For information on royal gardens, see Tony Spawforth's *Versailles* and James Meader's *Planter's Guide*. Also see the website for the Chateau de Versailles (http://en .chateauversailles.fr/homepage) and for Malmaison Chateau (http://musees-nationaux -malmaison.fr/chateau-malmaison).

For information on the use of flower oils for skin care and healing, see Colleen Dodt's *The Essential Oils Book*.

Quote

55 "The species has great charm . . . especially when fruiting." Lenz, *Bonsai from the Wild*.

CHAPTER 8: A VIRGINIA NATIVE

For excellent information on Jefferson's gardening practices, see Andrea Wulf's *Founding Gardeners* and the bibliography entry for Peter Loewer and for Edwin Betts and Hazlehurst Perkins.

Jefferson described his plans for the grounds at Monticello in his *Garden Book* and in his *Account Book 1771*, under "The Open Ground on the West—a shrubbery." Many excerpts from his records are found on the Monticello website: www.monticello.org. See also the bibliography entries for Philip Miller, Joan Dutton, and Ann Leighton.

For more on Jefferson's attempts to disprove Buffon's theory of degeneracy, see Lee Dugatkin's *Mr. Jefferson and the Giant Moose*.

Quote

64 "Though I am an old man . . . but a young gardener." Thomas Jefferson, letter to Charles Wilson Peale, August 20, 1811, quoted on the Monticello website blog *Memorandum*: http://blog.monticelloshop.org/?p=1346.

CHAPTER 9: THE VINE LIFESTYLE

For more on the complex inner workings of vines, see Stephen Mulkey et al.'s *Tropical Forest Plant Ecophysiology*.

Quotes

70 "To ascertain . . . confined by illness." Darwin, *Movements and Habits of Climbing Plants*.

72 "Vines have . . . a massive trunk." Ibid.

73 "It has . . . organised beings." Neve and Messenger, *Charles Darwin.*

CHAPTER 10: THE COLUMBIAN EXCHANGE

For information on the Columbian Exchange, see both bibliography entries for Charles Mann, as well as the books by Patricia Fara and Alfred Crosby.

John Josselyn, an English botanist, published *New-England's Rarities Discovered*, a survey of the flora and fauna of New England, in 1672. Josselyn distinguished native plants from "plants as have sprung up since the English Planted and kept Cattle in New-England."

For information on planting invasives, see the GPO report *Plant Materials for Conservation.*

For more information on invasive species, see Alan Burdick's *Out of Eden.*

Quotes

78 "The Columbian exchange . . . all unknowing." Mann, *1493.*

79 "Plants that . . . our detriment." John Bartram, quoted in Berkeley Jr. and Berkeley, *Life and Travels of John Bartram.*

CHAPTER 11: WHAT DO ANIMALS EAT?

For information on research into wildlife feeding habits, see Alexander Martin et al.'s classic *American Wildlife and Plants.* The introduction has much fascinating information on the history and ecology behind the statistics. Also see Donald Stokes's *Natural History of Wild Shrubs and Vines.*

Two online resources are likewise of note: the Biodiversity Database of the Washington, DC, Area (http://biodiversity.georgetown.edu/index.cfm) and the Rouge River Bird Observatory (www.rrbo.org), which has been investigating the importance of urban natural areas to birds since 1992.

CHAPTER 12: BIRD CANDY

For more information on bluebirds, see Lawrence Zelany's *The Bluebird: How You Can Help Its Fight for Survival.*

For more on birds' winter feeding habits, see the bibliography entries for Bernd Heinrich, David Allen Sibley, Donald and Lillian Stokes, and Gilbert Waldbauer.

Information on the fat content and digestibility of poison ivy is from Julie Craves, "Poison Ivy: Breakfast of Champions."

CHAPTER 13: PREPARING FOR DOOMSDAY

For information on the deliberate sensitization of guinea pigs using poison ivy leaves, see Ernest Stratton's case study in *California and Western Medicine.*

For more on plant evolution, see Wilson Stewart and Gar W. Rothwell's *Paleobotany and the Evolution of Plants.*

For a listing of insects observed feeding on poison ivy, see Dale Habeck's "Insects Associated with Poison Ivy and Their Potential as Biological Control Agents." See also BugGuide.net, a website of Iowa State University, which has amazing close-up photographs of insects and their food plants.

For insights on poison ivy's point of view, see David Attenborough's *The Private Life of Plants* and Michael Pollan's *The Botany of Desire*.

CHAPTER 14: HOLDING ON TO THE LAND

For information on the history of Truro, Massachusetts, see the town website (www .truro-ma.gov/about/pages/history) and the National Park Service's webpage on Highland Light Station (www.nps.gov/nr/travel/maritime/hig.htm).

For more on dune ecology, see the bibliography entries for Janine Benyus, Rachel Carson, and Dorothy Sterling.

For information on the growth habits of poison ivy, see Edward Frankel's *Poison Ivy, Poison Oak, Poison Sumac and Their Relatives*. This wonderful resource has a wealth of information about the *Toxicodendrons*.

Quotes

101 "What now remains ... being left." Plato, *Critias*, as translated in the Perseus Digital Library of Tufts University: www.perseus.tufts.edu/hopper/text?doc=Perseus%3Atext %3A1999.01.0180%3Atext%3DCriti.%3Apage%3D111.

101 "The nation ... destroys itself." Franklin D. Roosevelt, quoted in Churchman and Landa, *The Soil Underfoot*.

102 "Beach plants ... waves." Janine Benyus, *Field Guide to Wildlife Habitats*.

104 "Our natural dune seawall ... man always loses." Watson, "Protect the Natural Dune Seawall."

104 "This has the very useful benefit ... behind it." Ibid.

CHAPTER 15: THERE'S GOLD IN THE HILLS: POISON OAK

For information on the historic landscape of the American West, see *Old Spanish Trail*, by Leroy and Ann Hafen, and *Los Angeles: A Guide to the City and Its Environs*, produced by the Federal Writers' Project.

For more on poison oak and its range, see Edward Frankel's *Poison Ivy, Poison Oak, Poison Sumac and Their Relatives* as well as the Biota of North American Program website: http://bonap.net/NAPA/TaxonMaps/Genus/County/Toxicodendron.

For more on Native American poison oak uses, see Virgil Vogel's *American Indian Medicine* and Jan Timbrook's "Virtuous Herbs."

For a skeptical viewpoint on uses of poison oak, see Thomas E. Anderson's *The Poison Ivy, Oak and Sumac Book*.

For more on the environmental impacts of mining on the American West, see the bibliography entries for Carolyn Merchant, H. W. Brands, and Randall Rohe.

See the website of the Centers for Disease Control and Prevention for information on the effect of breathing smoke laden with urushiol: www.cdc.gov/niosh/topics/plants.

The US Forest Service maintains an online database of plants with much useful ecological information, including their value in controlling erosion and restoring soil, how the plants are affected by fire, and their interrelationships with wildlife. See www.fs.fed .us/database/feis/plants/vine/toxdiv/all.html.

For information on nurse plant research, see the bibliography entries for Hai Ren et al. and Stephanie Yelenik et al.

Quotes

106 "We entered . . . the river." Juan Crespí, quoted in Hafen and Hafen, *Old Spanish Trail.*
108 "The credulity . . . credit for." Anderson, *Poison Ivy, Oak and Sumac Book.*
110 "It is impossible . . . appalling." Rossiter W. Raymond, 1872, quoted in Bakken and Farrington, *Environmental Problems.*

CHAPTER 16: THE DEVIL YOU KNOW

For more information on glyphosate's aliases, see the website of the Monsanto Company: www.monsanto.com.

For more information on glyphosate, see "Glyphosate" at www.epa.gov/ingredients -used-pesticide-products/glyphosate.

For information on glyphosate as an enzyme suppressant and its effects on human health, see the bibliography entries for Anthony Samsel and Stephanie Seneff; Crystal Gammon; Christina Howe et al.; and Ted Steinberg.

For information on the risks of herbicides to nontarget flora and fauna, see Carissa Ganapathy's "Environmental Fate of Triclopyr" and the website of Californians for Alternatives to Toxics (http://alt2tox.org).

For information on the links between pesticide exposure and asthma, see Antonio Hernández et al.'s "Pesticides and Asthma."

The Audubon Society estimates that approximately seven million birds are killed by the use of lawn and garden pesticides. According to Dr. Ward Stone, who was for many years wildlife pathologist for the New York State Department of Environmental Conservation and a nationally recognized expert on wildlife pesticide mortality, our knowledge of pesticides' effects on birds is like a pyramid-shaped iceberg, in which the widest portion by far remains unseen. Only a few dead birds are noticed and collected, but it's probable that the majority of bird deaths caused by pesticides go undetected. See more information in this flyer disseminated by the Missouri River Bird Observatory: http://mrbo.org/downloads/pdfs/Audubon's%20LawnFlyer.pdf.

Quotes

117 "Negative impact . . . Alzheimer's disease." Samsel and Seneff, "Glyphosate's Suppression."
118 "Died or became immobile." "Selected Commercial Formulations of Triclopyr— Garlon 3A and Garlon 4 Risk Assessment Final Report," www.fs.fed.us/r5/hfqlg/ publications/herbicide_info/1996b_triclopyr.pdf.

119 "Infants and children . . . liver and kidneys." "Pesticides and Children," National Pesticide Information Center, http://npic.orst.edu/health/child.html.

CHAPTER 17: NO HOLDS BARRED

The website of the Brookline, Massachusetts, Parks and Open Space Division has excellent information about the value of open space and native species. It mentions that "poison ivy provides berries that are edible for birds and rodents, leaves that certain insects eat, and shelter for the many native animals that do not have an allergic reaction to it." See www.brooklinema.gov/672/Concern-with-Non-Native-Plants.

For information on battling poison ivy and many fascinating episodes of humans versus poison ivy, see the bibliography entries for Thomas Anderson, George Tapley, and Joe Lamp'l.

Much information on the effects of fire on poison ivy is found on the website of the US Forest Service. The USFS maintains an online database of plants with a great deal of useful ecological information. See www.fs.fed.us/database/feis/plants/shrub/toxspp/all.html#BotanicalAndEcologicalCharacteristics.

For information on copper sulfate, see the National Pesticide Information Center, a website administered by Oregon State University: http://npic.orst.edu/factsheets/cuso4gen.html.

For more on the effects of ammonium sulfamate, see EXTOXNET (Extension Toxicology Network), a pesticide information project sponsored by five US universities: http://extoxnet.orst.edu.

For more on the effects of sodium arsenite, see the National Center for Biotechnology Information's PubChem database: http://pubchem.ncbi.nlm.nih.gov/compound/sodium_arsenite#section=Safety-and-Hazards.

Quote
121 "A kind of landscape . . . wild shrubbery." Downing, *Treatise on the Theory and Practice of Landscape Gardening*, as quoted on the website of the town of Brookline, Massachusetts: www.brooklinema.gov/DocumentCenter/Home/View/2519.

CHAPTER 18: THE FUTURE OF POISON IVY

For an amazing view of one possible future of our planet, see Alan Weisman's *The World Without Us*.

Quotes
126 "Increasing CO_2 resulted . . . with increasing CO_2." Ziska et al., "Rising Atmospheric Carbon Dioxide."
128 "The news came . . . Thunder knew that it was his son." Mooney, *History, Myths, and Sacred Formulas*.

EPILOGUE

For more on wildlife-friendly habitats created by humans, see *The National Wildlife Federation's Guide to Gardening for Wildlife*, by Craig Tufts and Peter Loewer. The National Wildlife Federation was an early advocate of backyard and urban landscaping for wildlife, and it has many excellent resources on its website: www.nwf.org.

Sara Stein's *Noah's Garden* first made me realize how America's landscape has changed and how homeowners can make positive changes by planting native plants.

Each of the fifty states has its own Native Plant Society. See the American Horticultural Society's list at www.ahs.org/gardening-resources/societies-clubs-organizations/native-plant-societies.

Quotes

132 "In wildness . . . world." Henry David Thoreau, quoted in Schneider, *Thoreau's Sense of Place.*

132 "Travelled . . . in Concord." Thoreau, *Walden.*

132 "A primitive . . . outward civilization." Ibid.

APPENDIX: HOW TO AVOID, HEAL, OBLITERATE, AND COEXIST WITH POISON IVY

Part 1: How Not to Get Poison Ivy

Euell Gibbons tried many of his own remedies, and eating one poison ivy leaf a day worked for him. See his *Stalking the Healthful Herbs.*

The estimate of the number of micrograms it takes to cause a reaction is from W. P. Armstrong and W. L. Epstein, "Poison Oak: More Than Just Scratching the Surface."

For more information on poison ivy rashes, see the Mayo Clinic Diseases and Conditions database: www.mayoclinic.org/diseases-conditions/poison-ivy/basics/causes/con-20025866.

See also the Toxicology Data Network, part of the US National Library of Medicine under the National Institutes of Health: http://toxnet.nlm.nih.gov.

Part 2: How to Treat Poison Ivy

For more information on jewelweed, see the bibliography entries for B. J. Zink et al., Maud Grieve, Tom Brown, Thomas Anderson, and Susan Hauser. Anderson's comprehensive *Poison Ivy, Oak and Sumac Book* has much information on the medical impacts of poison ivy on humans and describes many helpful remedies. Hauser's excellent *Field Guide* contains a wealth of information on poison ivy.

For more on Native American traditional medicine, see the bibliography entries for James Herrick and C. Erichsen-Brown. (It is Herrick who includes the remedy of using the juice of one plug of chewing tobacco.) Also see the Native American Ethnobotany website maintained by the University of Michigan: http://herb.umd.umich.edu.

For more on the effects of poison ivy on the skin, see David Crosby's *The Poisoned Weed.*

Quote

142 "It's like . . . can get in." Lisa Shea, "Poison Ivy and Hot Showers" on Lisa Shea website: www.lisashea.com/birding/poisonivy/poisonivyhotshower.html.

Part 3: How to Get Rid of Poison Ivy

See the website of the New York State Department of Environmental Conservation for more information on integrated pest management, particularly the pamphlet "IPM In and Around Your Home": www.dec.ny.gov/docs/materials_minerals_pdf/pm2.pdf.

   Information on the percentage of protein in poison oak is from Thomas Anderson's *Poison Ivy, Oak and Sumac Book.*

Quote

149 "An emerging . . . sustainable control." Silliman et al., "Livestock as a Potential Biological Control Agent."

Bibliography

Books and Articles

Allen, William. "Plant Blindness." *BioScience* 53, no. 10 (2003). http://bioscience.oxford
journals.org/content/53/10/926.full.

Anderson, Thomas E. *The Poison Ivy, Oak and Sumac Book: A Short Natural History and Cautionary Account.* Ukiah, CA: Acton Circle, 1995.

Armstrong, W. P., and W. L. Epstein. "Poison Oak: More Than Just Scratching the Surface." *Herbalgram* 34 (1995): 36–42.

Atkinson, D. T. *Magic, Myth, and Medicine.* New York: Fawcett, 1956.

Attenborough, David. *The Private Life of Plants.* London: BBC Books, 1995.

Audubon, John James. *The Birds of America.* New York: Sterling, 2012. First published in 1837.

Audubon, Maria, and Elliot Coues, eds. *Audubon and His Journals.* New York: Charles Scribner's Sons, 1897.

Bakken, Gordon, and Brenda Farrington. *Environmental Problems in America's Garden of Eden.* London: Routledge Press, 2001.

Benyus, Janine. *The Field Guide to Wildlife Habitats.* New York: Simon and Schuster, 1989.

Berkeley Jr., Edmund, and Dorothy Smith Berkeley, eds. *The Correspondence of John Bartram, 1734–1777.* Gainesville: University Press of Florida, 1992.

Berkeley Jr., Edmund, and Dorothy Smith Berkeley. *The Life and Travels of John Bartram: From Lake Ontario to the River St. John.* Gainesville: University Press of Florida, 1990.

Betts, Edwin, and Hazlehurst Bolton Perkins. *Thomas Jefferson's Flower Garden at Monticello.* Charlottesville: University of Virginia Press, 1971.

Bigelow, Jacob. *American Medical Botany.* Vol. 3. Boston: Cummings and Hilliard, 1817.

Blackburn, Julia. *The Emperor's Last Island: A Journey to St. Helena.* New York: Vintage Books, 1993.

Blunt, Wilfrid. *The Compleat Naturalist: A Life of Linnaeus.* New York: Viking Press, 1971.

Boyd, Jane E., and Joseph Rucker. "No Ill Nature: The Surprising History and Science of Poison Ivy and Its Relatives." *Chemical Heritage* 31, no. 2 (Summer 2013). www.chemheritage.org/discover/media/magazine/articles/31-2-no-ill-nature.aspx?page=1.

Bradford, William. *Of Plymouth Plantation.* New York: Random House, 1981. First published in 1856 as *History of Plymouth Plantation.*

Brands, H. W. *The Age of Gold: The California Gold Rush and the New American Dream.* New York: Doubleday, 2002.

Brown, Tom. *Guide to Wild Edible and Medicinal Plants.* New York: Berkley Books, 1985.

Burdick, Alan. *Out of Eden: An Odyssey of Ecological Invasion.* New York: Farrar, Straus, and Giroux, 2005.

Caddy, Florence. *Through the Fields with Linnaeus: A Chapter in Swedish History.* London: Longmans, Green, 1887.

Carson, Rachel. *The Edge of the Sea.* New York: Houghton Mifflin, 1955.

Churchman, G. Jock, and Edward, R. Landa, eds. *The Soil Underfoot: Infinite Possibilities for a Finite Resource.* Boca Raton, FL: CRC Press, 2014.

Craves, Julie. "Poison Ivy: Breakfast of Champions." *Net Results,* blog of the Rouge River Bird Observatory, University of Michigan–Dearborn, October 19, 2011. http://net-results.blogspot.com/2011/10/poison-ivy-breakfast-of-champions.html.

Crosby, Alfred. *The Columbian Exchange: Biological and Cultural Consequences of 1492.* Santa Barbara, CA: Praeger Press, 2003. First published in 1973.

Crosby, David. *The Poisoned Weed.* Oxford: Oxford University Press, 2004.

Darwin, Charles, and Francis Darwin. *The Power of Movement in Plants.* Lenox, MA: Hardpress, 2010. First published in 1880.

———. *The Movements and Habits of Climbing Plants.* New York: New York University Press, 2010. First published in 1875.

Dodt, Colleen. *The Essential Oils Book.* Pownal, VT: Storey Books, 1996.

Downing, Andrew Jackson. *A Treatise on the Theory and Practice of Landscape Gardening.* Boston: C. C. Little, 1841.

Dufresnoy, Andre Ignace Joseph. *Des caractères, du traitement et de la cure des dartres: des convulsions . . . de la vomique . . . etc. Par l'usage du Rhus-radicans.* Paris: L'Imprimerie de Delance, 1788.

"Dufresnoy on Rhus." *British Journal of Homoeopathy* 28, no. 113 (1870).

Dugatkin, Lee. *Mr. Jefferson and the Giant Moose.* Chicago: University of Chicago Press, 2009.

Dutton, Joan Parry. *Plants of Colonial Williamsburg.* Williamsburg, VA: The Colonial Williamsburg Foundation, 1979.

Epstein, William L. "Allergic Contact Dermatitis." In *Current Perspectives in Immunodermatology,* edited by Rona M. MacKie. London: Churchill Livingstone, 1984.

———. "Occupational Poison Ivy and Oak Dermatitis." *Dermatologic Clinics* 4, no. 2 (April–June, 1986): 511–16.

Erichsen-Brown, C. *Medicinal and Other Uses of North American Plants.* New York: Dover, 1979.

Erikson, Carrolly. *Josephine: A Life of the Empress.* New York: St. Martin's, 2000.

Fara, Patricia. *Sex, Botany and Empire.* New York: Columbia University Press, 2004.

Farrington, Harvey. *Homoeopathy and Homoeopathic Prescribing.* New Delhi, India: Jain, 2001.

Ford, Alice, ed. *Audubon by Himself.* Garden City, NY: The Natural History Press, 1969.

Frankel, Edward. *Poison Ivy, Poison Oak, Poison Sumac and Their Relatives.* Pacific Grove, CA: Boxwood Press, 1991.

Gammon, Crystal. "Weed-Whacking Pesticide Proves Deadly to Human Cells." *Scientific American,* June 23, 2009. www.scientificamerican.com/article/weed-whacking -herbicide-p.

Ganapathy, Carissa. "Environmental Fate of Triclopyr." Report for Environmental Monitoring and Pest Management Branch, Department of Pesticide Regulation, Sacramento, CA. January 2, 1997. www.cdpr.ca.gov/docs/emon/pubs/fatememo/ triclopyr.pdf.

Gerard, John. *The Herball, or Generall Historie of Plantes.* New York: Dover, 1975. First published in 1597 and revised in 1633.

Gibbons, Euell. *Stalking the Healthful Herbs.* New York: David McKay, 1971.

———. *Stalking the Wild Asparagus.* Brattleboro, VT: Alan C. Hood, 1962.

Graham Jr., Frank. "Audubon's Legacy: Where It All Began." *Audubon,* Nov.–Dec. 2004.

Grieve, Maud. *A Modern Herbal.* New York: Dover, 1971. First published in 1931.

Habeck, Dale H. "Insects Associated with Poison Ivy and Their Potential as Biological Control Agents." *Proceedings of the VII International Symposium on the Biological Control of Weeds,* edited by Ernest S. Delfosse. Melbourne, Australia: CSIRO, 1989. www.invasive.org/proceedings/pdfs/7_329-337.pdf.

Hafen, Leroy R., and Ann Hafen. *Old Spanish Trail.* Lincoln: University of Nebraska Press, 1993.

Hauser, Susan Carol. *A Field Guide to Poison Ivy, Poison Oak, and Poison Sumac: Prevention and Remedies.* Guilford, CT: Globe Pequot Press, 2008.

Heinrich, Bernd. *Winter World: The Ingenuity of Animal Survival.* New York: Harper-Collins, 2003.

Hernández, Antonio F., Tesifón Parrón, and Raquel Alarcón. "Pesticides and Asthma." *Current Opinion in Allergy and Clinical Immunology* 11, no. 2 (2011): 90–96.

Herrick, James W. *Iroquois Medical Botany.* Syracuse: Syracuse University Press, 1995.

Hightshoe, Gary. *Native Trees, Shrubs, and Vines for Urban and Rural America.* New York: Van Nostrand Reinhold, 1988.

Hoffmann, Nancy, and John C. Van Horne. *America's Curious Botanist: A Tercentennial Reappraisal of John Bartram (1699–1777).* Philadelphia: American Philosophical Society, 2004.

Howe, Christina, Michael Berrill, Bruce D. Pauli, Caren C. Helbing, Kate Werry, and Nik Veldhoen. "Toxicity of Glyphosate-Based Pesticides to Four North American Frog Species." *Environmental Toxicology and Chemistry* 23, no. 8 (August 2004): 1928–38.

Johnson, William. *Cottage Gardener and Country Gentleman's Companion.* Vol. 11. London: William S. Orr and Co., 1854.

Jönsson, Ann-Mari. "The Reception of Linnaeus's Works in Germany with Particular Reference to His Conflict with Siegesbeck." Paper presented at the Conference on Germania Latina/Latinitas Teutonica, Munich, Germany, September 2001. www.phil-hum-ren.uni-muenchen.de/GermLat/Acta/Jonsson.htm.

Josselyn, John. *New-England's Rarities Discovered.* Boston: Massachusetts Historical Society, 1972. First published in London, 1672.

Keynes, Randal. *Darwin, His Daughter, and Human Evolution.* New York: Riverhead, 2002.

Laird, Mark. *The Flowering of the Landscape Garden: English Pleasure Grounds, 1720–1800.* Philadelphia: University of Pennsylvania Press, 1999.

Lamp'l, Joe. "The Plague of Poison Ivy." Wicked Local Brookline, *Brookline Tab*, January 29, 2013. http://brookline.wickedlocal.com/article/20130129/NEWS/301299902.

Leighton, Ann. *American Gardens in the Eighteenth Century: "For Use or for Delight."* Boston: Houghton Mifflin, 1976.

Leith-Ross, Prudence. *The John Tradescants: Gardeners to the Rose and Lily Queen.* London: Peter Owen, 1984.

Lenz, Nick. *Bonsai from the Wild: Collecting, Styling, and Caring for Bonsai.* Passumpsic, VT: Stone Lantern, 2007.

Linnaeus, Carolus. *A General System of Nature, Through the Three Grand Kingdoms of Animals, Vegetables, and Minerals; Systematically Divided into Their Several Classes, Orders, Genera, Species, and Varieties, with Their Habitations, Manners, Economy, Structure, and Peculiarities.* Edited by William Turton. London: Lackington, Allen, 1802. www.biodiversitylibrary.org/bibliography/37018.

———. *Philosophia Botanica.* Translated by Stephen Freer. New York: Oxford University Press, 2003. First published in 1751.

———. *Species Plantarum.* Laurentii Salvii, 1753. www.biodiversitylibrary.org/item/13829.

Linwood, Custalow, and Angela Daniel. *The True Story of Pocahontas: The Other Side of History.* Golden, CO: Fulcrum, 2007.

Loeffler, Amy. "Scientists Weed Out Pesky Poison Ivy with Discovery of Killer Fungus." *Science Newsline*, June 12, 2014. www.sciencenewsline.com/summary/2014061219030063.html.

Loewer, Peter. *Jefferson's Garden.* Mechanicsburg, PA: Stackpole Books, 2004.

Mann, Charles C. *1491: New Revelations of the Americas before Columbus.* New York: Vintage Books, 2005.

———. *1493: Uncovering the New World Columbus Created.* New York: Alfred A. Knopf, 2011.

Martin, Alexander C., Herbert S. Zim, and Arnold L. Nelson. *American Wildlife and Plants: A Guide to Wildlife Food Habits.* New York: Dover, 1951.

Meader, James. *The Planter's Guide, or Pleasure Gardener's Companion.* London: G. Robinson, 1779.

Merchant, Carolyn. *Green versus Gold: Sources in California's Environmental History.* Washington, DC: Island Press, 1998.

McLynn, Frank. *Napoleon: A Biography.* New York: Arcade, 1997.

Miller, Philip. *The Gardener's Dictionary: Containing the Best and Newest Methods of Cultivating and Improving the Kitchen, Fruit, Flower Gardens, and Nursery; as Also for*

Performing the Practical Parts of Agriculture, Including the Management of Vineyards, with the Methods of Making and Preserving Wine, According to the Present Practice of the Most Skillful Vignerons in the Several Wine Countries in Europe: Together with Directions for Propagating and Improving, from Real Practice and Experience, All Sorts of Timber Trees. London: John and Francis Rivington, 1768.

Millspaugh, Charles F. *American Medicinal Plants.* New York: Dover, 1974. First published in 1892.

Mohan, J. E., L. H. Ziska, W. H. Schlesinger, R. B. Thomas, R. C. Sicher, K. George, and J. S. Clark. "Biomass and Toxicity Responses of Poison Ivy (*Toxicodendron radicans*) to Elevated Atmospheric CO_2." *Proceedings of the 2006 National Academy of Sciences* 103, no. 24 (2006): 9086–89.

Mooney, James. *History, Myths, and Sacred Formulas of the Cherokee.* Fairview, NC: Bright Mountain Books, 1992. First published in 1891.

Mulkey, Stephen, Robin L. Chazdon, and Alan P. Smith. *Tropical Forest Plant Ecophysiology.* New York: Springer, 1996.

Neve, Michael, and Sharon Messenger, eds. *Charles Darwin.* Autobiographies series. New York: Penguin Classics, 2006.

Parkinson, John. *Theatrum Botanicum: The Theater of Plants: Or, An Herball of Large Extent: Containing Therein a More Ample and Exact History and Declaration of the Physicall Herbs and Plants . . . Distributed Into Sundry Classes Or Tribes, for the More Easie Knowledge of the Many Herbes of One Nature and Property.* London: Tho. Cotes, 1640.

Pavord, Anna. *Naming of Names: The Search for Order in the World of Plants.* New York: Bloomsbury, 2005.

Peattie, Donald Culross. *A Natural History of Trees of Eastern and Central North America.* Boston: Houghton Mifflin, 1948.

Percy, George. "Discourse of the Plantation of the Southern Colony in Virginia by the English, 1606." In *Narratives of Early Virginia, 1606–1625,* edited by Lyon Gardiner Taylor. New York: Barnes & Noble, 1959. First published in 1907. https://archive.org/stream/narrativesofearl1946tyle/narrativesofearl1946tyle_djvu.txt.

Phillips, Charles D. F. *Materia Medica and Therapeutics of the Vegetable Kingdom.* New York: William Wood, 1879.

Piffard, Henry G. *A Treatise on the Materia Medica and Therapeutics of the Skin.* New York: William Wood, 1881.

Plant Materials for Conservation. Washington, DC: Government Printing Office, 1979.

Platter, Thomas, and Horatio Busino. *The Journals of Two Travellers in Elizabethan and Early Stuart England.* Canton, NY: Caliban Press, 1994.

Pollan, Michael. *The Botany of Desire.* New York: Random House, 2002.

Potter, Jennifer. *Strange Blooms: The Curious Lives and Adventures of the John Tradescants.* London: Atlantic, 2007.

Price, David. *Love and Hate in Jamestown: John Smith, Pocahontas, and the Start of a New Nation.* New York: Random House, 2003.

Quammen, David. *The Reluctant Mr. Darwin: An Intimate Portrait of Charles Darwin and the Making of His Theory of Evolution.* New York: W. W. Norton, 2007.

Ren, Hai, Long Yang, and Nan Liu. "Nurse Plant Theory and Its Application in Ecological Restoration in Lower Subtropics of China." *Progress in Natural Science* 18, no. 2 (February 2008): 137–42. www.sciencedirect.com/science/article/pii/S1002007107000202.

Rohe, Randall. "Man and the Land: Mining's Impact in the Far West." *Arizona and the West* 28 (Winter 1986): 299–338.

Rubin, Leslie. "Selected Commercial Formulations of Triclopyr—Garlon 3A and Garlon 4 Risk Assessment Final Report." www.fs.fed.us/r5/hfqlg/publications/herbicide_info/1996b_triclopyr.pdf.

Samsel, Anthony, and Stephanie Seneff. "Glyphosate's Suppression of Cytochrome P450 Enzymes and Amino Acid Biosynthesis by the Gut Microbiome: Pathways to Modern Diseases." *Entropy* 15, no. 4 (April 2013): 1416–63. www.mdpi.com/1099-4300/15/4/1416.

Schmidt, Susan. *Landfall along the Chesapeake: In the Wake of Captain John Smith.* Baltimore: Johns Hopkins University Press, 2006.

Schneider, Richard J., ed. *Thoreau's Sense of Place: Essays in American Environmental Writing.* Iowa City: University of Iowa Press, 2000.

Scudder, John Milton. *Specific Medication and Specific Medicines.* Cincinnati: Baldwin, 1870.

Sibley, David Allen. *The Sibley Guide to Bird Life and Behavior.* New York: Knopf, 2001.

Silliman, Brian R., Thomas Mozdzer, Christine Angelini, Jennifer E. Brundage, Peter Esselink, Jan P. Bakker, Keryn B. Gedan, et al. "Livestock as a Potential Biological Control Agent for an Invasive Wetland Plant." *PeerJ*, September 23, 2014. https://peerj.com/articles/567.

Slaughter, Thomas P. *The Natures of John and William Bartram.* Philadelphia: University of Pennsylvania Press, 1996.

Small, E., P. M. Catling, J. Cayouette, and B. Brookes. *Audubon: Beyond Birds: Plant Portraits and Conservation Heritage of John James Audubon.* Ottawa, Ontario: Canadian Science, 2009.

Smith, John. *The Complete Works of Captain John Smith (1580–1631).* Edited by Philip L. Barbour. 3 vols. Chapel Hill: University of North Carolina Press, 1986. www.virtualjamestown.org/johnsmith.html.

Souder, William. *Under a Wild Sky: John James Audubon and the Making of* The Birds of America. New York: North Point, 2004.

Spawforth, Tony. *Versailles: A Biography of a Palace.* New York: St. Martin's, 2008.

Stein, Sara. *Noah's Garden: Restoring the Ecology of Our Own Backyards.* New York: Houghton Mifflin, 1993.

Steinberg, Ted. *American Green: The Obsessive Quest for the Perfect Lawn.* New York: W. W. Norton, 2006.

Sterling, Dorothy. *The Outer Lands.* New York: W. W. Norton, 1978.

Stewart, Wilson N., and Gar W. Rothwell. *Paleobotany and the Evolution of Plants.* Cambridge: Cambridge University Press, 1993.

Stoever, Dietrich Heinrich. *The Life of Sir Charles Linnaeus, Knight of the Swedish Order of the Polar Star.* Translated by Joseph Trapp. London: B. and J. White, 1794.

Stokes, Donald W. *The Natural History of Wild Shrubs and Vines*. New York: Harper and Row, 1981.

Stokes, Donald, and Lillian Stokes. *The Stokes Field Guide to the Birds of North America*. New York: Little, Brown, 2010.

Stratton, Ernest K. "Poison Oak and Poison Ivy Immunity Studies in Guinea Pigs." *California and Western Medicine* 54, no. 3 (March 1941): 115–16. www.ncbi.nlm .nih.gov/pmc/articles/PMC1634037/pdf/calwestmed00295-0017.pdf.

Tantaquidgeon, Gladys. *Folk Medicine of the Delaware & Related Algonkian Indians*. Philadelphia: Philadelphia Historical Commission, 2000. First published in 1940.

———. *A Study of Delaware Indian Medicine Practice and Folk Beliefs*. University of Pittsburgh Library system. First published in 1942. https://archive.org/details/ studyofdelawarei00tant.

Tapley, George. "Cities Can Destroy Poison Ivy: Brookline Massachusetts Has Found an Effective, Safe Method." *American City*, August 1944: 53.

Taylor, Lyon Gardiner, ed. *Narratives of Early Virginia*. New York: Charles Scribner's Sons, 1907.

Thomas, Joseph. *The Universal Dictionary of Biography and Mythology*. New York: Cosimo Classics, 2013.

Thoreau, Henry David. *Walden*. New York: Signet Classics, 2012. First published in 1854.

Timbrook, Jan. "Virtuous Herbs: Plants in Chumash Medicine." *Journal of Ethnobiology* 7, no. 2 (Winter 1987): 171–80. http://ethnobiology.org/sites/default/files/pdfs/ JoE/7-2/Timbrook1987.pdf.

Tufts, Craig, and Peter Loewer. *The National Wildlife Federation's Guide to Gardening for Wildlife: How to Create a Beautiful Backyard Habitat for Birds, Butterflies, and Other Wildlife*. Emmaus, PA: Rodale Press, 1995.

US Environmental Protection Agency. "Children Are at Greater Risks from Pesticide Exposure." Pesticide Fact Sheet, January 2002. http://archive.epa.gov/pesticides/ regulating/laws/fqpa/web/html/kidpesticide.html.

Vogel, Virgil. *American Indian Medicine*. Norman: University of Oklahoma Press, 1970.

Wagner, David. *Caterpillars of Eastern North America*. Princeton, NJ: Princeton University Press, 2005.

Waldbauer, Gilbert. *The Birder's Bug Book*. Cambridge: Harvard University Press, 1998.

Watson, Richard L. "Protect the Natural Dune Seawall and Prevent Hurricane Destruction at Port Aransas, Texas." Texas Coastal Geology website, October 2005. http://texascoastgeology.com/beach.pdf.

Weisman, Alan. *The World Without Us*. New York: St. Martin's, 2007.

Wilson, Edward O. *Biological Diversity: The Oldest Human Heritage*. Albany: New York State Biodiversity Research Institute, 1999.

Writers' Program, Works Progress Administration. *Los Angeles: A Guide to the City and Its Environs*. New York: Hastings House, 1941.

Wulf, Andrea. *The Brother Gardeners*. New York: Random House, 2008.

———. *Founding Gardeners*. New York: Random House, 2011.

Yelenik, Stephanie G., Nicole DiManno, and Carla M. D'Antonio. "Evaluating Nurse Plants for Restoring Native Woody Species to Degraded Subtropical Woodlands." *Ecology and Evolution* 5, no. 2 (January 2015): 300–13. www.ncbi.nlm .nih.gov/pmc/articles/PMC4314263.

Zelany, Lawrence. *The Bluebird: How You Can Help Its Fight for Survival.* Bloomington: Indiana University Press, 1978.

Zink, B. J., E. J. Otten, M. Rosenthal, and B. Singal. "The Effect of Jewel Weed in Preventing Poison Ivy Dermatitis." *Journal of Wilderness Medicine* 2, no. 3 (August 1991): 178–82.

Ziska, L. H., R. C. Sicher, K. George, and J. E. Mohan. "Rising Atmospheric Carbon Dioxide and Potential Impacts on the Growth and Toxicity of Poison Ivy (*Toxicodendron radicans*)." *Weed Science* 55, no. 4 (July–August 2007): 288–92.

Zweig, Stefan. *Marie Antoinette.* New York: Viking, 1954.

WEBSITES

The Algonquin Way: www.thealgonquinway.ca

American Academy of Dermatology: www.aad.org

American Horticultural Society: www.ahs.org

Ashmolean Museum: www.ashmolean.org

Audubon's *Birds of America*, University of Pittsburgh Digital Research Library: http:// digital.library.pitt.edu/a/audubon

Bartram's Garden: www.bartramsgarden.org

Biodiversity Database of the Washington, DC, Area: http://biodiversity.georgetown .edu/index.cfm

Biota of North America Program: www.bonap.org

Brookline, Massachusetts: www.brooklinema.gov

BugGuide: http://bugguide.net

Californians for Alternatives to Toxics: http://alt2tox.org

Chateau de Versailles: http://en.chateauversailles.fr

Encyclopedia Virginia: www.encyclopediavirginia.org

Extension Toxicology Network: http://extoxnet.orst.edu

George Clifford Herbarium: www.george-clifford.nl/UK/hc_UK.htm

Highland Light Station, Truro, Massachusetts: www.nps.gov/nr/travel/maritime/hig .htm

Homeopathy for Health: www.homeopathyforhealth.net

Malmaison Chateau: http://en.musees-nationaux-malmaison.fr/chateau-malmaison

Mayo Clinic: www.mayoclinic.org

Mohan Research Lab: http://mohanlab.uga.edu

Monsanto Company: www.monsanto.com

National Center for Homeopathy: www.nationalcenterforhomeopathy.org

National Institute for Occupational Safety and Health: www.cdc.gov/niosh/topics/ plants

National Pesticide Information Center: http://npic.orst.edu

National Wildlife Federation: www.nwf.org

Native American Ethnobotany Database: http://herb.umd.umich.edu
Natural History Museum, London: www.nhm.ac.uk
New York Flora Atlas: www.newyork.plantatlas.usf.edu
Rouge River Bird Observatory: www.rrbo.org
Swedish Museum of Natural History: www.nrm.se
Texas Coastal Geology: http://texascoastgeology.com
Truro, Massachusetts: www.truro-ma.gov/about/pages/history
USDA Forest Service, Fire Effects Information System: www.feis-crs.org/beta
Virtual Jamestown: www.virtualjamestown.org

Index

About the Author

Anita Sanchez is a science writer fascinated by plants and animals no one loves—like dandelions, tarantulas, and poison ivy. She worked as an environmental educator for over twenty-five years at the New York State Department of Environmental Conservation, leading outdoor classes on poison ivy–lined trails. Her blog, www.unmowed.com, sings the praises of unnoticed and unappreciated plants.